Howard

# THE
# CIVILIZED DEFENSE PLAN

# THE CIVILIZED DEFENSE PLAN

## Security of Nations Through the Power of Trade

By Howard S. Brembeck
The Fourth Freedom Forum

**HERO BOOKS**
Fairfax, Virginia

ISBN 0-915979-21-7

Printed and bound in the USA

HERO BOOKS
10392 Democracy Lane
Fairfax, Virginia  22030
703-491-3674
FAX 703-591-6109

and

Distributed in Great Britain
1990
by
Greenhill Books, Lionel Leventhal Limited,
Park House, 1 Russell Gardens,
London NW11  9NN

ISBN 1-85367-056-1

# About the Author

Howard S. Brembeck is an inventor and industrialist with a life-long commitment to the worldwide improvement of agricultural efficiency—a commitment that has given him the global insights that led to his innovative plan for global preservation.

The poultry and livestock feeding and watering equipment pioneered by Brembeck has played a significant role in improving feed utilization and holding down the price of farm animal products on every continent. He has been issued a number of patents and the company he founded, CTB, Inc., is the largest of its kind in the world, doing business in more than 80 nations.

Born in Wabash, Indiana, and raised on a farm, Brembeck built his successful business from scratch and he continues to serve as chairman of the board of CTB with headquarters in Milford, Indiana. He has stayed close to his roots in America's heartland but, in the tradition of fellow Hoosier Wendell L. Willkie, he long has perceived the reality of today's one world. He has opened production facilities or established joint ventures in a number of countries, gaining first hand experience of the world's economic, political and social interdependence. At the same time, he has intensively studied international relations, giving special attention to the roles of the United States and the Soviet Union.

Brembeck and his wife, Myra, have one daughter, two grand-children and two great grand-children. His family knows from experience that he enjoys, to use his words, "creating something that didn't exist before." His breakthrough developments have played a role in the fight to achieve freedom from want and now his persistence and creativity have targeted another global challenge, achieving freedom from fear.

# Dedication

To the people, living and dead, of all nations who have given of themselves so that all of us may someday live in a truly civilized world free from the fear of war.

To Lawrence Burkholder whose long ago encouragement started me on the quest that led to this book.

Most of all, to my wife, Myra. Without her unfailing support and non-complaining assistance there probably never would have been a Civilized Defense Plan.

# Contents

# Foreword

On October 6, 1983, this newspaper headline commanded my attention: "INDUSTRIALIST SEES WORLD TRADE AS ALTERNATIVE TO ARMS POWER." The industrialist was Howard Brembeck and his idea so intrigued me that I contacted him and, a few months later, I became a founding member of what is now the Fourth Freedom Forum.

Since then, I have been joined by many others and, although I cannot presume to know all of their motives, I suspect that they, too, were attracted by what they perceived to be an innovative approach to international security. I also suspect that they shared with me feelings of fear and frustration generated by a series of books, films and television productions that vividly portrayed the horrors of a nuclear holocaust. These materials attracted many readers and viewers and they played a major role in raising world consciousness of the magnitude of the problem. But for the most part, they failed to offer a credible solution.

When I first became aware of Howard Brembeck and his ideas, I had come to realize that it is one thing to describe and decry the ills of the world; it is quite another to create a cure. It became apparent to me that the arms race was an unprecedented problem that conventional approaches had failed to solve and that we had to recognize what innovators have known for centuries: Unprecedented problems demand unprecedented solutions.

So when a businessman from America's heartland refused to point the finger of blame and started designing a practical solution to the problem, he struck a chord that reverberated within me and many others. It is a chord that now is beginning to make itself heard throughout the world.

Just a few years ago, the ideas described by Howard Brembeck in this book would have been scoffed at, passed off as ridiculous and looked upon in the same light as Don Quixote's Impossible Dream. But "impossible" only means that it hasn't happened yet and, when you look at the record of the past few years, you see that the impossible is starting to happen. Gorbachev and his new thinking came to power in the Soviet Union. History's first U.S.-U.S.S.R. arms reduction agreement was completed and carried out. The Soviets admitted that their economy was coming apart at the seams and the United States suddenly became the largest debtor nation in history. At the same time, Germany and Japan gained recognition as prototypes of a new kind of superpower, one based in economic rather than military prowess.

Because of these developments, Howard Brembeck's ideas are more than timely. They are a prophetic revelation of the world as it rapidly is becoming. In this world, security is defined in economic, social and ecological as well as military terms. It is a world that needs to direct a greater share of its resources to the reduction of a variety of threats to the continuing well-being of the human race and, at the same time, to provide an adequate, effective defense against the aggressive behavior of unscrupulous national leaders.

In today's world, change is occurring at a dizzying pace. We have an obligation to do our best to make sure change works in the best interest of our generation and the generations to come. It is time to sieze the opportunities change is creating and to rise to Howard Brembeck's challenge to create a Civilized Defense Plan to protect our future and our world.

*Marc A. Hardy, Executive Director*
*Fourth Freedom Forum*

# Acknowledgements

I owe a debt of gratitude to the many individuals—and I can name more than 100—who, in one way or another, helped develop the Civilized Defense Plan presented in this book. To avoid the risk of inadvertently overlooking anyone, I have chosen not to attempt to list all of them and to ask that they accept my sincere appreciation for the comments, suggestions, insights and opinions that helped shape my thinking.

I do want to name a special few who made important, specific contributions: Lawrence Burkholder and Andrew Hardie, who for nearly 10 years kept urging me on; Charles Ainlay, who was instrumental in helping me refine the Plan's Premise V, making it a more positive statement of the Plan's power to achieve lasting international security; Jonas Salk, who advised me on how to approach a supposedly unsolvable problem; Robert Schuller and Milton Friedman, who took the time to give me a critical review of the Plan during its early stages, and Lloyd Dumas, Ben Ferencz, Carla Johnston, Senator Richard Lugar, Congressmen Jim Courter and John Hiler, John Gilligan and others who never hesitated to present their candid opinions of my ideas. I also am indebted to the men and women who participated in the Fourth Freedom Forum's 1988 and 1989 Roundtable sessions. The constructive criticism of the Plan brought forward by these individuals resulted in the strengthening of a number of sections.

Over a period of many years, some of the most important suggestions for Plan refinement came from other dedicated directors, advisers and associates, past and present, of the Fourth Freedom Forum: Dorothy Ainlay, Kay Ball, Phillip Berry, Harold Chestnut, Christopher Chocola, Michael Closson, James Evans, John Frieden, Marc Hardy, Jan Harshberger, William Johnson, Ian Mitroff, Rod Morris, Miriam Redsecker, James Tierney, Paul Walker, and Burns Weston as well as the Fourth Freedom Forum's administrative assistant, Ann Miller. I extend my thanks to them and to James Carroll, my editor, word massager and thought contributor without whom the Plan may not have made it into book form.

I also owe much to my professor brothers, Winston and Cole Brembeck, who relentlessly and mercilessly critiqued drafts of this book and who still may not be completely happy with it even though it incorporates much of their advice. Finally, I thank my daughter, Caryl, her husband, Byron, my granddaughter, Kelly, and grandson, Chris, for never appearing to doubt my sanity though I know they questioned why a businessman like me would think he could devise a plan that could save the world from self-destruction.

*Howard S. Brembeck*

# Introduction

Let me begin with a word of explanation and warning. This is a book about the world as it is now and as it can be in the future. This is a book about a defense that can protect us from weapons of mass destruction, military aggression and state-sponsored terrorism. The book describes a new approach to security. It calls for the establishment of a common defense for all the world's people, the building of a system that will lift the burden of fear, the creation of an environment in which humanity can achieve its potential. The book explores a route to international security that differs fundamentally from the present failed approach. If you are unwilling to suspend conventional patterns of thought, to look at the international security issue from a different vantage point, reading further could be a waste of your time.

If you do decide to read on, as I hope you will, I ask that you do so with a critical eye. If you have a criticism, please pass it on to me. I promise to give your thoughts careful attention. Indeed, criticisms from many individuals in many areas of expertise contributed greatly to the development of the ideas the book describes. It also should be noted that the book is neither a scholarly treatise nor a neatly packaged tour of a landscape we already have seen through the eyes of politicians and pundits. It is an invitation to join me in a journey of discovery, one that will follow a new pathway of thinking about defense. We will visit a landscape that has been hidden from view. There you will see with your own eyes opportunities for living at a higher level, free from the fear of weapons of mass destruction, armed aggression and the covert form of aggression called terrorism. This invitation is extended to all who are concerned about the awesome threats now facing civilzation—including the skeptic who does not believe freedom from fear is possible and the thoughtful individual who has considered the problem and, out of despair, has made a subconscious decision to deal with history's greatest challenges by ignoring them.

My own pioneering of this hidden landscape began many years ago. I long have had a deep interest in international affairs and, as the founder and chairman of a multinational corporation doing business in more than 80 nations, I have had ample opportunity to observe U.S. foreign policy and its results. Like most Americans, I tended over the years to put my faith in the wisdom, experience and insight of the policy makers in the White House, State Department and Congress. I'll stick to my business, I thought, and let the experts in Washington stick to theirs. Sometimes I felt encouraged and optimistic. More often, I felt that many of our policies were clearly wrong-headed, ill-advised, self-defeating and out of touch with reality. I was especially discouraged when I considered the foreign affairs area in which I had a special interest: relations between the United States and the Soviet Union.

This interest of mine goes back as far as 1942 when I read a book by Dr. Harry Rimmer entitled *The Coming War and the Rise of Russia* (William B. Eerdmans Publishing Company, Grand Rapids, Michigan, 1940). At the time that book was written, Germany and the Soviet Union had a non-aggression treaty and Hitler seemed in complete control of continental Europe. Yet, Dr. Rimmer was predicting that Germany would be defeated and the Soviet Union would become a dominant world force. Dr. Rimmer also predicted that the Balkan nations would come under Soviet control and Libya and Ethiopia would become Soviet clients. As many of these predictions started to come true, I made an increasing effort to penetrate the cloud of mystery and contradiction that so often seems to surround international events and the policies that spawn them.

Much of my attention was concentrated on the Soviet Union, a nation that seemed the epitome of mystery and contradiction. My interest bordered on obsession and, for nearly 40 years, I read a wide range of materials on the Soviet Union and the philosophy and thought processes of its leaders. I also sought out emigres from the Soviet Union and questioned them about the thoughts, concerns and dreams of the Soviet people. As the Soviet Union's power increased, so did my admiration for Dr. Rimmer. His ability to understand the significance of international developments was demonstrated by predictions that were proving more accurate than I had ever imagined. The Soviet Union had become a superpower and it was locked in an ever-escalating arms race with the United States. I was convinced that the giants were on a collision course. War could only be averted through the utilization of some powerful and, in my mind, unknown mechanism.

The search for such a mechanism was never far from my thoughts. Then while riding on a bus in England the concept of the Civilized Defense Plan took shape in my mind. *It was recognition of the most obvious, the simple fact that the United States had the responsibility as well as the ability to take the lead in solving the problem of nuclear arms and other weapons of mass destruction. But the United States would have to act out of world interest, not self-interest, in joining with other nations to pull the world back from the brink of self-annihilation.* The other nations of the world—industrial and pre-industrial—lack the power to provide the leadership that can move society beyond reliance on military force. On the other hand, the United States is the world's sole economic and military superpower. It is uniquely equipped to work with other nations in steering the world in a new direction. In addition, the United States has a society that can adapt to the new ideas required to solve the most pressing and dangerous of new problems.

This is an appropriate role for the United States. The nation that first created the atomic bomb should assume responsiblity for returning to the people of the earth a world free from weapons of mass destruction. This is not a matter of guilt. There is every reason to believe that those who built the bomb and those who used it were acting in good conscience. They wanted to quickly end the most devastating war in world history and prevent the bloodbath that would have been an inevitable result of an invasion of the Japanese homeland. These were very desirable objectives. Nonetheless, by developing and employing the bomb, the United States took the lead in the creation of an enormous new problem for the world. And those who create problems have a responsibility to solve them. Indeed, I have found that to create a problem and not correct it often creates much greater problems.

But what can the United States do now to change the course of history? I am as convinced as I was when I first conceived the Civilized Defense Plan (CDP) that the answer is unmistakable: *Prudent use of economic power.* Through economic incentives and sanctions, utilized with uniformity and without malice, the Civilized Defense Plan offers benefits to all nations. It offers a pathway to a world dedicated to mutual economic development, a world free of weapons of mass destruction. The Civilized Defense Plan offers a way for the world to avoid the ever present possibility of catastrophic conflict.

The Civilized Defense Plan name is entirely accurate. It is based on a definition which states that *civilization is a way of living that respects the right of others to live*. This is the fundamental difference between CDP and other systems of defense. Instead of relying on destructive physical force, CDP draws its strength from something much more versatile and persuasive. On the positive side, CDP employs economic power to reward nations that commit to the elimination of weapons of mass destruction. On the negative side, CDP employs economic power to penalize nations that refuse to make such a commitment. Through this positive-negative inducement, CDP can create a defense that actually defends, without sacrificing the lives of countless civilians.

CDP, which now has been in the process of development for 10 years, is described in some detail on the following pages. You will also find a recounting of some of the thinking that led to the formulation of the Plan. However, I believe it is important to begin with these points:

1.   This book is devoted to explaining a concept that has the power to free the world from fear of weapons of war and other devastating threats to society. The book does not waste time decrying our present precarious situation. It sees digging up the past, which usually results in accusations and excuses, as an exercise in futility and an obstacle to effective problem-solving. At the same time, CDP does not ignore the question of responsibility. It is based on the belief that if we want to end the antagonisms that lead to violent behavior, we must start with ourselves. We must be willing to objectively evaluate the potential reaction to all of our actions. We must always assume that when we have a problem, we are at least partly to blame.

2.   The natural law of compensation—the "absolute balance of give and take" described by Ralph Waldo Emerson—helps us understand why violence inevitably leads to more violence, why war always begets more war. In today's world, armed conflict is unmistakably counterproductive. Technology has created a one-world society. The notion of nations relating to each other in terms of physical force has become entirely and obviously obsolete. *Nations must do as their citizens have done: Submit to the rule of law*.

3.   Like many people in nations on both sides of the East-West dividing line, I tend to be suspicious of the

trustworthiness of Soviet leaders. But while my suspicions are evident in this book, I want to make it clear that when it comes to untrustworthy aggressive behavior, no power, major or minor, has an unblemished record. In the absence of enforceable law, all nations have been and are potential offenders.

4. While my initial concern was with the threat of nuclear weapons, the book's discussion of CDP must be understood to include all the barbaric instruments of aggression. In a resolution dated August 12, 1948, the United Nations Commission for Conventional Armaments defined weapons of mass destruction as including ". . . atomic explosive weapons, radioactive material weapons, lethal chemical and biological weapons, and any weapons developed in the future which have characteristics comparable in destructive effect to those weapons . . . mentioned above." This definition has been accepted by all nations.

5. When I began to develop the CDP idea in 1979, my principal focus was on the nuclear and other potentially catastrophic weapons that have dominated the strategic thinking of the world's great powers for decades. The book reflects this focus. However, I also am suggesting that CDP can be applied to the solution of other complex international problems. So the book includes a limited discussion of how CDP can be used to free the world from the threat of state-sponsored terrorism.

6. Civilization threatening weapons are too dangerous to leave in the hands of the experts—the politicians, diplomats and military leaders who created the present threat to earth and its people. Individuals must take charge of their destiny and start giving direction to their officials.

7. The pacifist movement has alerted the world to the threat posed by nuclear weapons. But CDP is not an outgrowth of this movement or of what is usually thought of as traditional pacifist thinking.

8. *CDP is based on a belief that nations have a sovereign right to defend themselves. They have no sovereign right to the weapons of mass destruction that threaten to annihilate the civilian population of other nations.* It also should be noted that they have no sovereign right to provide sanctuaries for terrorists.

9. CDP is not pro-capitalist or pro-communist or pro any other system of social, political or economic thought. CDP is impartial and objective. It is biased solely toward the survival of this generation and those to come.

10. Implementation of CDP undoubtedly will be opposed by at least some elements of the military, by weapons producers, by

some politicians and by others who may think they benefit from preservation of the status quo. They must be shown that CDP offers even greater benefits. It will increase international security, enhance global trade and contribute to the well-being of all nations.

11.   CDP is not intended to be a final, detailed, step-by-step blueprint for a new order of peaceful human society. CDP is not an end in itself. It is a means of achieving an objective. It is a tool that can be used to bring about the establishment of enforceable international law. CDP is a positive approach, a concept, a catalyst that can help change the way we think about our world and its future.

12.   Unlike some of the visionary peace proposals that attract attention from time to time, CDP does not seek to remake men and women. It accepts them as they are. It assumes that they will strive to achieve higher ethical and behavioral standards but it doesn't expect the upward progress of the next 100 years to greatly exceed that of the past 100 years. CDP is based on a realistic view of humanity as it is, not on the wishful belief that exhortation and aspiration can soon produce a new and improved model.

Most of all, the Civilized Defense Plan is a vision of what is within our power to achieve, an ideal that can become real, a tomorrow that we can bring into existence. So let us begin by taking an initial look at CDP itself. I ask you to remember that while it gives priority attention to the elimination of nuclear and other civilian-threatening weapons, it can have other applications. Indeed, CDP's basic concept might be applied very effectively to the terrorism problem.

I say "very effectively" because civilized nations do not tolerate terrorism and they are frustrated and embarrassed by their inability to deal with the handful of nations that employ terrorists to commit acts of armed aggression against civilians. On the other hand, many nations that consider themselves civilized tolerate the most uncivilized of weapons and they are not at all embarrassed by their addiction to the doctrine of mass destruction. So in this book, I concentrate most of my attention on the most challenging problem of our age—nuclear and similarly dangerous weapons.

Now I ask you to join me in a journey of exploration. The starting point is CDP's foundation, the five basic premises that deal with weapons of mass destruction, outline CDP's scope and dimensions and reveal why CDP is an unconventional but thoroughly practical approach to international security.

*Howard  S. Brembeck*

# CIVILIZATION

A way of living that respects the

right of others to live

# TRADE

A power that built civilizations

A power that can build a civilized world

# The Plan For A Civilized Defense: Five Fundamental Premises

## I

The United States has a responsibility to take the lead in a multinational effort to free the world from weapons which constitute a massive threat to civilian life, weapons which drain human and economic resources, contribute to a climate of international distrust and diminish the hope that earth's people can work together to meet common challenges and achieve common goals.

## II

As the keystone of today's interconnected and interdependent world economy, the United States has the influence to enlist the support of other nations in the development of a world law that would eliminate weapons of mass destruction. Support would be obtained through economic incentives and, if necessary, the imposition of sanctions on trade in manufactured goods. To insure the effectiveness of these sanctions and in sharp contrast with the sanctions of the past, they would apply uniformly to nations unwilling to support the law *and to nations that continue to trade with nations not supporting the law*.

1

## III

A world law would go into effect when it is adopted (1) by two-thirds of all the world's nations and (2) by nations doing two-thirds or more of all international commerce in manufactured goods. This formula, which must include the Soviet Union, assures persuasive political and economic support for the law and its objectives.

## IV

For enforcement of the law banning weapons of mass destruction, supporting nations would establish a limited purpose international agency with the authority and ability to define, detect and dispose of such weapons presently in existence, prevent further manufacture of such weapons, identify acts of armed aggression, monitor trade with outlaw nations, marshal world opinion against aggressors and issue the call for imposition of economic sanctions.

## V

Support for international law would bring to a nation all the economic, technical, social and other benefits of world trade. Violators of international law and their accomplices would forfeit these benefits and be subjected to complete economic isolation and worldwide public scorn.

# *PREMISE I*

## Underlying Concepts
## And
## U.S. Responsibilities

The United States has a responsibility to take the lead in a multinational effort to free the world from weapons which constitute a massive threat to civilian life, weapons which drain human and economic resources, contribute to a climate of international distrust and diminish the hope that earth's people can work together to meet common challenges and achieve common goals.

# 1

# Demilitarizing Our Concept of Power

*The unleashed power of the atom has changed everything save our modes of thinking, and thus we drift toward unparalleled catastrophe. We shall require a substantially new manner of thinking if mankind is to survive.*

— Albert Einstein

If you watch the nightly television news or read the newspaper editorials, you know the world is experiencing what only can be described as a revival of hope. The Reagan-Gorbachev summits, the INF Treaty, the continuation of the Strategic Arms Reduction Talks, the sweeping disarmament proposals of President Gorbachev and President Bush and the unexpected outbreak of peace in the most unlikely locations have been taken as indications that the ancient problem of international security at last can be solved.

Now hope certainly is a desirable virtue. It helps all of us achieve our goals by assuring us that the goals are attainable. But when you take a hard look at the current situation, when you measure it against Einstein's familiar call for "a substantially new manner of thinking," you cannot help but wonder why hope is back in style. Many of the political leaders of nations large and small do not seem to understand the currents of change that have transformed and are continuing to transform the world. In area after area of urgent concern, some world leaders seem blind to reality, unable to comprehend that some of their most cherished beliefs about power have been rendered obsolete by time, trade and technology.

Nowhere is this obsolescence more evident than in the world's

5

continued reliance on military force. During the past 20 years, the greatest military powers, the United States and the Soviet Union, failed miserably when they attempted to use military force to achieve national objectives. After sacrificing thousands of lives and billions of dollars, the United States withdrew from Vietnam. The Soviet Union, apparently learning nothing from the experience of its superpower rival, followed much the same scenario in Afghanistan. Smaller nations are not immune to this form of blindness, either. Iraq attacked Iran when that nation supposedly was preoccupied with other problems, provoking a winnerless, marathon war that may prove to be the bloodiest since World War II. Israel and its Arab neighbors have been in a nearly perpetual state of war and neither side has achieved its announced goals of justice and security. No matter where you look in today's world, armed conflict produces not victory but frustration and defeat.

Why is armed force no longer an effective instrument of national policy? Perhaps one reason is the way defensive technology has managed to keep pace with offensive technology. For example, in Afghanistan the introduction of relatively inexpensive Stinger missiles gave a Mujahideen foot soldier the power to destroy a helicopter gunship worth millions of rubles. Only in nuclear and, possibly, chemical-biological weapons does the offense have a clear advantage, for reasons that are as obvious as they are frightening. But if technology has failed to neutralize the most devastating weapons, it has succeeded in creating a world in which increasing numbers of people understand that the use of military power is an outmoded concept. Through television, war has been brought into the living room. Ordinary citizens have seen what many of their leaders refuse to see—that armed force is counterproductive, that it yields only pain, death and the hatred that foments continuing conflict. The powerful images transmitted around the world by television are creating a climate of public opinion that will not tolerate military adventures. Television certainly played a role in America's withdrawal from Vietnam. World opinion, mobilized by television, undoubtedly helped convince Soviet leaders of the wisdom of marching home from Afghanistan. Should the Soviet army again intervene in Hungary or Czechoslovakia, as it did in the 1950s and '60s, the action would provoke worldwide outrage, perhaps extending into the Soviet Union itself.

For decades, technology has been quietly erasing the boundaries that traditionally separate nations. Barriers have been broken down, creating for us a new world society or, more accurately, a one world society. We can talk to people almost any place on earth in a matter of minutes and, through a relatively inexpensive electronic device, we can transmit a document across the continent or a lunch order to the delicatessen down the street. In a matter of hours, we can travel almost any place on earth. We can visit Europe or Asia more easily than America's founding fathers could visit a neighboring colony. Through the news media, Americans and the citizens of other nations with a free press witness distant events and we have the sensation of being physically present at the events we witness. The telephone, the airplane, television and the computer—which contributed to every recent technological advance, including the conquest of space—have brought the people of the world together in ways that were unforeseen and perhaps unforeseeable a generation ago.

> **. . . technology has built a global society and, at the same time, technology has dismantled the traditional means of insuring global security.**

In one of history's countless ironies, technology has built a global society and, at the same time, technology has dismantled the traditional means of insuring global security. We are neighbors in a world community but the police no longer are effective and we have no sure way of protecting our lives and property from the world's burglars and bandits. We can put locks on our doors and windows. But we have no way of getting the criminals off the streets. The frustration felt by civilized society is evident. For example, witness the powerlessness of the world's great powers when it comes to eliminating terrorism. Armies of security officers appear as ineffective against the terrorist as the Maginot Line was against the blitzkrieg. Of course, a great power could use military force to attack the source of the problem. But every military action is by nature provocative. It carries with it the possibility that it will trigger a wider conflict, a danger that most national leaders are unwilling to risk in this nuclear age.

Yet, the leaders of nations continue to cling to the military myth. They build their arsenals because, when they look back into history, they see that relationships within the world community always have been conditioned by the existence of usable military force. So it continues to be a key ingredient in the foreign policy considerations of every nation. If only the world's leaders would look at the present and into the future, they might become aware of the dwindling effectiveness of military action. They might begin to understand that the use of military force has proven to be neither productive nor acceptable. They might see that the world is confronted by a security vacuum, one that can be filled if only we decide to change the way we think about power.

To begin with, power is not inherently evil. In itself, power is neither good nor evil. But it is a necessary element for the functioning of any civilized society. It is wishful thinking to believe that international security can be achieved without employing power. But to achieve good, power must be kept under control. Gasoline burning in the open is wasteful and potentially destructive. Gasoline burning in the controlled environment of an engine yields power that can be harnessed to produce all kinds of positive results. We also have to realize that physical force and power are not the same. They do not equate. Physical force, which includes military force, is only one of many kinds of power. *In fact, there are two kinds of power that are far more persuasive than physical force: Economic power and the power of withholding.* These two powers—used with prudence, resolution and control—can achieve goals that never can be achieved by military force.

Economic power and the power of withholding are universal in character, politically and ideologically neutral, eminently practical and, most important of all, they provide for our physical needs and serve as primary tools for the improvement of our condition. Used together, these two forms of power can diminish the threat of global destruction and enforce the rule of law in international affairs. The power of economics and the power to withhold have the ability to free the world's people from fear of the awesome weapons that have been accumulated in a fruitless pursuit of international security. The power of economics and the power to withhold can induce nations to accept the rule of law and punish the criminal nations that disrupt our world community. The power of economics and the power to withhold can be used in more situations than can military force and they are far less likely to create the kind of a violent reaction that is an inescapable result of military action.

The power to withhold is so frequently used, so natural, so much a part of our daily lives that we rarely acknowledge its importance. In a way, it is like the sun. We know the sun exists but we don't stop to think of it as the source of the energy that fuels our earth. We know that withholding exists, but we don't stop to consider that our ability to withhold influences our transactions with others, creating the conditions that shape our lives. Individuals, organizations and governments use the power to withhold every day to effect change, improve conditions, exercise control or gain an advantage. A parent withholds privileges from an unruly child. Consumers withhold purchases to influence the behavior of manufacturers or retailers. Employees withhold their labor from employers to improve working conditions or wages. Governments withhold money, privileges and freedom to obtain compliance with their laws and maintain their ascendency. Both physical force and withholding can be used to improve a situation. But there is a profound difference in the way they are used and in the effect they have on those who use them and on those they are used against.

Inherent in the use of physical force is the reaction of those against whom it is used, a reaction directed against the users. This reaction can be so strong and so violent that it eliminates all the gains that force was supposed to achieve. As we know from history, war sows the seeds of war, creating hostilities that extend from generation to generation.

> **Withholding achieves the desired results because implied in its use is the ability to give. Physical force totally lacks this ability. It can take away but it cannot give.**

On the other hand, withholding is nonviolent and rarely produces a violent reaction. Withholding achieves the desired results because implied in its use is the ability to give. Physical force totally lacks this ability. It can take away but it cannot give. The power to withhold is effective because it has positive as well as negative attributes. It works because it involves the possibility of transferring something desirable from one party to another. It incorporates both the carrot and the stick. Physical force, on the other hand, is one-dimensional. It is entirely negative. It is incapable of providing rewards. It offers no carrot, only a stick. If the stick fails to produce results, physical force has only one answer: Get a bigger stick.

The ability to give, inherent in the power to withhold, makes possible successful agreements, treaties and laws. All of these are, in essence, contracts between two or more parties. A contract is successful when it benefits all parties that are involved. A contract that does this will endure. A contract that is unfair because it favors one party over another—a likely occurrence when a contract is negotiated at gunpoint—is almost certain to create more problems than benefits, even for the favored party. The party that is less favored almost always will seek to violate the agreement by any means, fair or foul. Because so many of the world's leaders—and private citizens as well—seem to be locked into outmoded and crippling habits of thought, the power of withholding and the power of economics rarely have been used effectively as substitutes for military power. When we think about power in the context of international affairs, military force continues to come to mind. We customarily rank nations according to their military capabilities. We speak of powers and superpowers in accordance with the number of divisions they have in the field and the size of their nuclear and conventional arsenals. We see evidence of a violent world and we are skeptical about nonviolent solutions to the world's problems.

Part of this skepticism is based on our perception of what has, or has not, happened when nonviolent solutions have been attempted. For example, the United States, the world's greatest economic power, has been frustrated in many of its attempts to exercise nonviolent power in pursuit of what it perceives as its national objectives. One of the reasons for the frustration is the flawed nature of U.S. efforts to utilize its economic resources to influence the course of international events. The United States has lacked a consistent and comprehensive plan. It has tended to utilize its economic power and the related power to withhold in an uncontrolled way. It sometimes has attempted to bring its power to bear on those who violate world order but it has not extended its effort to include those who trade with the violators. It has withheld trade from violators but not from their accomplices. As a result, the violators can shrug their shoulders at a U.S. embargo. They know they can find another source of the goods they need.

If the United States and other nations are serious about utilizing nonviolent power in the international arena, the act of withholding cannot be limited to a single nation. To achieve the results we desire, trade also must be withheld from nations that trade with those violating international order. If the United States, in concert with other nations, really wants to achieve lasting international security, *trade must be withheld from nations that are threats to security and from their trading partners as well.*

We also need to adjust some of our thinking about the U.S. role in world affairs. We have to understand that self-interest and world interest are not always identical. We must realize that we can be most effective when we put aside self-interest and act in behalf of the interests of the world community. In the long run, this broader vision will pay major dividends to the United States. It will create unprecedented opportunities for our human and material resources. It will enable us to exercise constructive leadership in a society that has been granted new hope of achieving its potential. It's also worth noting that the United States, like many other nations, has tended to emphasize sanctions rather than inducements, punishment rather than rewards. Americans and their leaders seem to forget that some of the greatest triumphs of U.S. foreign policy came when we utilized our ability to give.

For example, after World War II, we gave our wealth in the rebuilding of Western Europe, including the nation that had been our principal European enemy. In so doing, we achieved a major foreign policy objective by creating an obstacle that would serve to contain Soviet expansion. We also created a robust trading partner, one that has been a major market for our goods and services, a source of products for the industrial and consumer segments of our economy and a competitive force that has stimulated innovation and productivity in American business. The lessons learned from the Marshall Plan and from the enlightened policies that contributed to the rebuilding of Japan can be applied again in the present moment of history. After World War II, we helped nations that were defeated militarily to get on their feet again. Now we have an opportunity to help nations that have been defeated economically. Then as now, we have much to gain. Then as now we can put a price on our assistance. In the post-World War II period, we demanded that our defeated military opponents renounce aggression and the weapons that make aggression possible. Today we cannot make demands on the Soviet Union and the other nations that have been defeated economically, but we can offer our assistance for a fair price. That price is support for an international weapons law that protects the people of all nations from weapons of mass destruction. In essence, this is the Civilized Defense Plan. Like the Marshall Plan, CDP is designed to eliminate emnity and aggression, to foster understanding, cooperation and mutual concern.

As the world's largest trading nation and with allies that also are among the leaders in world trade, the United States has a special opportunity and a special responsibility. It is in a unique position to use its economic power for good. Through our ability to withhold trade—to reward as well as persuade—we can enlist the support of our trading partners. We can mobilize a majority of the world's nations in building the political concensus that is the first step in the creation of an international law that will effectively protect all people from weapons of mass destruction. It is a law that will be welcomed by most nations, a law that will work because it will meet the test of fairness and provide equal protection to all nations.

Should the United States and other nations fail to use their nonviolent power in this way, should their economic power be wasted on narrow  self-interest and their energies misspent on unenforceable arms agreements, the opportunity to create lasting and genuine international security could be lost forever. The leadership responsibilities of the United States are at the heart of the Civilized Defense Plan. By facing up to the reality of the 20th Century and the part the United States has played in shaping this period of human history, Americans can exert a powerful positive influence on the world and its future. In cooperation with friends and allies from around the globe, they can usher in a truly civilized age. To give you an initial idea of how this can be accomplished, the next chapter takes a preliminary look at the Plan itself.

# 2

# A Civilized Approach
# To Defense

*I wish to see the discovery of a plan that would induce
and oblige nations to settle their disputes without cutting
one another's throats.*

—Benjamin Franklin

When we look back through time, World War II emerges as the
dreadful moment in history when the fragile boundary between
soldier and civilian was forever obliterated. It was a war that
made prime targets of the very young and the very old, a war that
seemed to have emerged from the most sinister recesses of the
human soul. Much of the world's population was terrified, not
only by the threat of destruction but by the fear that humanity had
lost its way and was blundering into a new and frightening age of
savagery. So when President Franklin D. Roosevelt spoke to Congress about what he called the Four Freedoms, he hit a responsive
chord. In that 1941 speech, he revived the hopes of millions. He
gave them the dream they needed for survival, a world in which
all people would enjoy freedom of speech and religion, freedom
from want and fear. Nearly a half century has passed since the
Four Freedoms were proclaimed and the dream is still a dream.
Freedom of speech and religion quite obviously are not enjoyed
by all people in all nations. Freedom from want remains beyond
the reach of vast segments of the earth's population. Worst of all,
freedom from fear seems less possible now than it did during the
most horrifying days of World War II.

The nuclear nightmare has taken up residence in our minds and

13

hearts. Fear has become an integral part of the human condition. Fear influences the conduct of nations and individuals, stifles global cooperation, shrinks our faith in our ability to control our own destiny and interferes with the quest for the freedoms that once seemed so close to attainment. Yet the dream of the Four Freedoms has not lost its appeal or validity. It remains a compellingly attractive goal because it deals with unchanging human concerns.

When President Roosevelt spoke of the Fourth Freedom, he could not have foreseen the incredible proliferation of nuclear weapons nor the computerized accuracy of the delivery systems that make every location a potential target. He spoke not of military hardware but of fundamental goals:

> *The fourth is freedom from fear, which translated into world terms, means a world-wide reduction of armaments, to such a point and in such thorough fashion that no nation will be in a position to commit an act of physical aggression against any neighbor, anywhere in the world.*

The reasoning is apparent. If the weapons of aggression no longer are permitted, military force and the threat of its use no longer will be an alternative available to national leaders. War will cease to be, as 19th century Prussian military strategist Karl von Clausewitz described it, "an instrument of national policy." Without weapons of aggression, nations will have to abandon all thought of imposing their wills on others with physical force. Then, free from fear, the people of the world will be on the threshold of a new age of progress. But how can weapons of mass destruction be eliminated? How can the tools of aggression be removed from the hands of the politicians and warriors of all nations? How can we move from the questionable security of Mutually Assured Destruction and its current derivatives to the genuine safety of Mutually Assured Survival? The answer is enforceable international law, a global compact which unites nations in a binding agreement to abolish nuclear and all other weapons of mass destruction.

There is nothing new about the concept of peace through international law. What is new is the understanding that this concept has become an imperative for the human race. As Albert Einstein put it, ". . . world authority and an eventual world state are not just desirable in the name of brotherhood, they are necessary for survival." Einstein was realistic when he used the word "eventual" in referring to a world state. The nations of earth do not appear ready to hand over their sovereignty to a global government. Indeed, cogent arguments can be raised against the radical central-ization of political authority. But the elimination of weapons of mass destruction does not require such centralization. It requires only that nations accept the sovereignty of a law dealing with the issue of life or death, the law that is a principal objective of the Civilized Defense Plan.

At every step in the development of government—from tribe to feudal society to city-state to nation—people have had to forfeit a portion of their sovereignty in order to achieve a benefit that is available only to those who are part of a larger political, social or cooperative relationship. Sometimes the benefit is protection from enemies. Sometimes, it is economic opportunity. With CDP, nations will forfeit their outdated perception of sovereignty over weapons that could destroy other nations. In return, they will achieve both protection from these weapons and increased eco-nomic opportunity. As outlined in the five basic premises, CDP calls for the United States to take a leadership role, working in cooperation with other nations and utilizing economic power to bring about creation of an international law prohibiting produc-tion, possession, sale and use of nuclear and all other weapons of mass destruction. The law would be enforced through rewards and penalties based on international trade and augmented by the strength of world opinion.

As the premises make clear, the law would go into effect when it is adopted by two-thirds of the world's nations doing at least two-thirds of all international commerce, including the Soviet Union. A nation supporting the law would be assuring itself of major economic benefits. It also would be agreeing to stop all trade with any nation that refuses to sign. The prohibition against trade would apply as well to nations that continue to do business with those nations that stay outside the law. In other words, both outlaw nations and their accomplices would suffer the consequences of a transgression. By targeting both transgressors and their accomplices, CDP takes a step beyond the flawed and ineffective sanctions of the past. It establishes a comprehensive system for the nonviolent enforcement of a law outlawing weapons of mass destruction. With the Civilized Defense Plan, the identity of the outlaw nations would be unmistakable: They would be the ones that refuse to ratify the international weapons law. It would be equally easy to identify the nations dedicated to the elimination of the weapons that contribute to aggressive national behavior.

These nations would be signatories of the law, nations acting in concert to build a world of expanded economic opportunity and enforceable international security.

The economic sanctions called for in the Civilized Defense Plan involve two distinct phases. Phase I would cover the period prior to the effective date of the law. During this period, nations supporting the law would impose economic sanctions on any nation that rejects the law. These phase I economic sanctions would be limited to trade in manufactured goods, the segment of the economy most directly involved in the production of weapons of mass destruction. For humane reasons, trade in agricultural and other earth products would be exempt. Also exempt would be medicines and related supplies and equipment.

Phase II economic sanctions would become operational after the law is approved by two-thirds of the nations doing two-thirds of the world's trade in manufactured goods and is declared binding on the world community. In this phase, the ban on trade would be all-inclusive. It would include earth products as well as manufactured goods. Both phase I and phase II economic sanctions are based on a recognition of the interdependence of all nations. In this age of interconnected economies and expanding technology, no nation can afford to go it alone, to isolate itself from two-thirds of the world's most highly industrialized nations. Any government that isolates its people from the international economy would be inviting its demise. Unlike the failed trade sanctions of the past, CDP advances the interests of all countries and groups of countries. It benefits all nations by recognizing that all people have a self-evident right to the Fourth Freedom.

---

**CDP proposes to replace the nuclear deterrent, which provides no defense, with the rule of law, which is both deterrent and defense.**

---

When you first examine CDP, even in the outline form presented in this chapter, you can see that it is outside the mainstream of conventional thought. It calls on people to look at their world in a different way, to put aside stereotypes of national interest and obsolete concepts of military force. CDP proposes to replace the nuclear deterrent, which provides no defense, with the rule of law, which is both deterrent and defense. With the rule of law in effect among nations, they will have the kind of defense now enjoyed by people who live under the rule of law within a nation.  CDP will create a world in which the lives of people in all nations will be respected.  As you further analyze the Civilized Defense Plan, you can see that it is profoundly practical.  It is based on a realistic view of a world that has experienced—and is still experiencing—fundamental change.  To understand these changes, to see where we are at this moment in history and where we may be headed, it's worth spending time reviewing where we have been.

# 3

# Transcending
# The Insanity
# Of Our Age

*Anybody who feels at ease in the world today is a fool.*
— Robert Maynard Hutchins

It's ironic—and well worth remembering—that the madness of the nuclear arms confrontation can be traced back to fear of a madman. The World War II leaders of the United States and Britain were afraid Hitler might get the bomb. They were certain he would not hesitate to use it to destroy London and, sooner or later, other population centers in Europe and America. The response of the allied leaders was predictable. They mobilized the scientific community in an all-out effort to beat Hitler to the prize, a weapon with unprecedented capacity for destruction. There seems to have been minimal discussion of the morality of nuclear warfare or of the long-range consequences of the introduction of such a powerful and little understood force. The bomb seems to have been viewed as merely another step in the deadly progression from club to crossbow to gunpowder and beyond. As President Harry S. Truman put it in his *Memoirs* (Doubleday & Company, Garden City, New York, 1955): "I regarded the bomb as a military weapon and never had any doubt that it should be used." Besides, the secret of the atomic bomb was in U.S. hands and why should anyone in the world be afraid of a country that was, at least in the view of Americans, so benign, so lacking in aggressive intentions and territorial ambitions?

19

What was largely overlooked by Americans, was how other nations perceived the bomb. Some saw it as a new destabilizing element in the equation of international relations. Many felt possession of nuclear weapons was the only way to assure an adequate national defense, to guard against the real or potential threats of others. Driven by fear or national pride, more and more nations joined the nuclear club and as the club's membership increased, so did overall military spending.

Nations large and small, rich and poor, spent increasing shares of their economic resources on weapons. *State of the World—1986* (W. W. Norton & Company, New York, New York, 1986), a report by the Worldwatch Institute written by the Institute's president, Lester R. Brown, and others, describes this phenomenon as "the militarization of the world economy."

In the most recent volume in the series, *State of the World—1989* (W. W. Norton & Company, New York, New York, 1989), the authors state:

> The world has spent an estimated 16 trillion dollars for military purposes since World War II. Adjusted for inflation, industrial countries have doubled their outlays since 1960 while third-world nations have increased their expenditures for military purposes six-fold. These numbers underestimate the real extent of militarization because they do not include militias and other para-military forces, and because there are no estimates for employment in arms industries in some nations. Perhaps most important though, is the burgeoning arms trade worth an estimated 635 billion dollars in 1984.
>
> Throughout history, governments have sought to develop and acquire more numerous and effective arms. Every major scientific discovery has been checked for possible military application. Possession of the perfect weapon, it was (and still is) believed would mean absolute security—unquestionable superiority, and thus leverage over other nations. Ironically, the nation that pioneered nuclear weapons development is now more vulnerable than ever. In less than 30 minutes, a single MX missile, or its Soviet counterpart, can deliver a destructive force equivalent to more than 200 Hiroshima bombs to within 90 meters of a target 11,000 kilometers away. The quest for the ultimate weapon has delivered the global community to an all encompassing state of insecurity.

There are senior military people in the United States and other nations who realize the insecurity created by nuclear weapons. They understand that these devices are not weapons of war but threats to human existence that should not be included in the arsenal of any nation. Unfortunately, most governmental leaders are so power oriented that they cannot comprehend this truth. They cannot see that nuclear weapons present a greater peril to humanity than AIDS and all other diseases combined. So when these leaders speak, we must be aware that they may not be speaking for us. We must not assume that the will of the people is the same as the will of their government. We must give our priority attention to the elimination of the weapons that have created this age of growing global insecurity.

By far the biggest players in the high stakes game that has been played with such negative results are the United States and the Soviet Union, the superpowers that have been locked in confrontation since the waning moments of the second World War. While estimates vary somewhat, it is quite generally believed that in recent years the United States has been devoting something less than seven percent of its gross national product (GNP) to defense. In fact, U.S. defense outlays for the fiscal year ended September 30, 1989, totaled $296.7 billion, or 5.7 percent of GNP. The Soviets, striving to achieve parity despite their much smaller economy, long were thought to have been devoting at least 14 percent of their GNP to defense. But the Soviets provided virtually no budget figures until May, 1989, when President Gorbachev told his nation's new parliament, the Congress of People's Deputies, that 1989 military spending totaled 77.3 rubles ($128 billion), or about 9 percent of GNP, a disclosure which may, or may not, offer a more reliable way to compare the military budgets of the two superpowers.

But while there continues to be disagreement on the relative size of military spending by the United States and the Soviet Union, there is general agreement that the long-standing military buildup has had a serious impact on the economies of both countries. Some respected economists believe the military spending of the United States has played a major role in several negative economic developments. For example, the federal government's debt has been climbing dramatically. The nation's traditional world trade surplus disappeared and the annual trade deficit climbed past the $150 billion mark. The United States, once the lender to the world, has become the largest debtor nation, not only in the world, but also in history. In the Soviet Union, both the industrial and agricultural segments of the economy rarely have been able to meet civilian demand. Although the Soviet economy is claimed to have experienced strong growth from time to time, its overall record in recent years has been dismal with the average annual rate of growth a lackluster two percent.

While a cause and effect relationship is not always provable, arms expenditures obviously are a heavy burden for the economies of both the United States and the Soviet Union. This certainly has provided a major incentive for the reduction of the stockpile of nuclear weapons in the possession of the United States and the Soviet Union. Both President Gorbachev and President Reagan realized that the arms race has inhibited economic development. They moved to eliminate all U.S. and Soviet intermediate range nuclear missiles and to seriously discuss a 50 percent reduction in long range strategic weapons, a move that appears to have the endorsement of the administration of President Bush. The President also has proposed substantial reductions in conventional weapons, a position that is at least somewhat in agreement with the troop reduction suggestions of President Gorbachev. Now all these are important and praiseworthy initiatives, but they clearly do not constitute even the beginning of a comprehensive system of international security. They do not eliminate from the minds of policymakers the doctrine of Mutually Assured Destruction— and such variations and derivatives as "strategic counterforce", "proportional response" and "flexible response".

---

**The fear of a madman that ushered in the atomic age has been replaced by wishful thinking that madness has been forever banned from the corridors of national power.**

---

The fear of a madman that ushered in the atomic age has been replaced by wishful thinking that madness has been forever banned from the corridors of national power. The bilateral arms agreements that have been achieved or proposed ignore the possibility of irrational behavior in the White House or the Kremlin. They do not deal with the reality of political, ideological or religious fanaticism. They cannot reassure those who worry that nuclear or other weapons of mass destruction will become increasingly available, that they will fall into the hands of renegade states or terrorist groups. Achieved and proposed arms agreements also give insufficient attention to the possibility of technological or human failure. Policymakers often seem to assume that the weapons of annihilation and the people who control them are fool-proof, that they will never malfunction or make mistakes—an assumption of perfection that is unlikely to be endorsed by many citizens of Norway who, in April, 1989, learned that a Soviet nuclear powered submarine armed with nuclear torpedoes caught fire and went to the bottom of the Norwegian Sea. Indeed, according to a study published in June, 1989, by Greenpeace, an environmentalist organization, and the Institute for Policy Studies, almost 1,300 major naval accidents occurred between 1945 and 1988, leaving on the ocean floor 48 nuclear warheads and nine nuclear reactors.

Anyone who objectively examines the concept of deterrence, and the arms agreements related to this concept, is bound to conclude that conventional diplomacy has built a shaky foundation for the future of humanity. At best, the world's leaders have temporarily delayed the nuclear onslaught. At worst, they have created a distracting illusion that has kept humanity from concentrating on a lasting, inclusive solution to the threat posed by nuclear weapons. This threat is not secret to anyone. A full-scale nuclear war would end life as we know it and almost no one believes a nuclear war could be anything less than full-scale for very long. As former Senator John Culver once said, attempting to limit the use of nuclear weapons is like "limiting the mission of a match thrown into a keg of gunpowder." When people face up to the facts of nuclear peril, they are horrified—but not for long. The ordinary reaction is to file the information away, to deny the existence of overwhelming destructive power on the specious grounds that "there's nothing I can do about it." This nuclear denial is one of the least realized and most dangerous aspects of the present MAD-created nuclear stalemate. Child psychiatrists have accumulated considerable evidence that many young people do not expect to reach adulthood. They believe they will be killed in a nuclear war. They are, as the April, 1984, *Psychology Today* noted in a headline, "Growing up scared . . . of not growing up." What weighs so heavily on the mind of a 10-year-old can weigh even more heavily on the mind of an adult. The result can be a fatalism that is not far removed from moral paralysis. Now is the time for the paralysis to end. We must confront the nuclear threat, recognizing its terrible potential. J. Robert Oppenheimer, one of the bomb's developers, summed it up after the successful testing in 1945 of a bomb that, by today's standards, was puny. Oppenheimer quoted from a Hindu text, the *Bhagavadgita*, "the radiance of a thousand suns . . . I am become as death, the destroyer of worlds."

The challenge all of us face is to exchange death for life, renounce the debilitating cult of fatalism and assume responsibility for our own future. As M. Scott Peck, M.D., put it in his bestselling book *The Road Less Traveled* (Touchstone Books, Simon and Schuster, New York, New York 1978):

> We cannot solve life's problems except by solving them. This statement may seem tautological or self-evident, yet it is seemingly beyond the comprehension of much of the human race. This is because we must accept responsibility for a problem before we can solve it. We cannot solve a problem by saying "It's not my problem." We cannot solve a problem by hoping that someone else will solve it for us. I can solve a problem only when I say "This is MY problem and it's up to me to solve it."

The nuclear threat is our problem. To solve it, we will have to break the habits of the past and begin to travel new roads to lasting security. Along these roads we will learn more about the thinking that is the foundation of CDP's Premise I, including the futility of unenforceable arms agreements and the necessity of enforceable international law.

# 4

# Seizing An Unprecedented Opportunity

*The dogmas of the quiet past are inadequate to the stormy present. We must think anew and act anew.*

— Abraham Lincoln

During the nearly 10 years of CDP development, one of the continuing assumptions was that it would be vigorously opposed by the Soviet Union. Such opposition seemed totally predictable, entirely in keeping with the record of Soviet hostility to proposals emanating from its superpower rival. Then came the Moscow summit and, soon afterward, the All-Union Communist Party Conference and multi-candidate elections. For the first time in memory, Soviet leadership seemed ready for change, anxious to embrace new ideas that would end political and economic stagnation. Mikhail Gorbachev and others talked openly of a new role for public opinion and the necessity of intensified competition.

To my surprise—and delight—Gorbachev and his followers seem to have reached first base in their effort to satisfy the rising expectations of the Soviet people. Home plate remains a distant and difficult goal. But the fact that such a goal exists, that Gorbachev is committed to transform many long-standing patterns of thought and behavior, is a development of major positive significance. If Gorbachev means what he says about openness and revitalization, the Soviet Union well could become one of the first nations to endorse CDP. By accepting the authority of an international law eliminating weapons of mass destruction, the Soviets

25

could free themselves from fear of devastating attack, reduce the drain of armaments expenditures and gain access to unrestricted trade with the United States and other industrialized democracies. By giving up the dubious advantage of maintaining an arsenal of civilian-threatening weapons, the Soviet people could enjoy the benefits of full participation in the enriching mainstream of international trade.

With so many advantages for East as well as West, CDP opens a new door to lasting and genuine international security. It relies on the rule of law to achieve an objective that has eluded the world—despite the best efforts of the able and well intended people who have devoted years to frustrating arms negotiations. These negotiations between the United States and the Soviet Union have failed to make the world safer. Indeed, the world is less safe now than it was before negotiations began. Each new round of arms limitation talks has been followed by the introduction of more and smarter weapons. We have gone from the bomber-dropped A-bomb of 1945 to multiple warheads on missiles that can be launched from the hidden depths of the ocean. This kind of development should come as no surprise. It is as old as human history. It reflects the obvious fact: *when people meet to negotiate, they seek not to weaken their relative position, but to improve it*. Negotiators for the United States and the Soviet Union represent constituencies that will not tolerate any concession that could be interpreted as a retreat. They are too wary of each other, too doubtful of the other's intentions. As a result, arms limitation talks have been most useful as platforms for propaganda, battlegrounds for the cold war and incubators for hatching new generations of more destructive weapons.

In fairness, it also must be conceded that there has been some progress in some areas of concern. The Limited Test Ban Treaty which prohibits nuclear test explosions in the atmosphere, outer space or underwater has been accepted by the United States, the Soviet Union and Great Britain. The Nuclear Nonproliferation Treaty is in operation and seems to have had a laudable effect on limiting the spread of material that could be utilized in the production of nuclear weapons. An entire class of weapons— intermediate range nuclear missiles—was eliminated by the treaty signed at the Moscow summit. In addition, Soviet and U.S. negotiators continue to discuss a number of other arms reduction issues that could have a positive, long term influence on international affairs. These discussions are well worth continuing. They could lead to a further reversal of the arms race, to the elimination of more of the weapons that threaten the future of earth and its people. The INF treaty and its implementation by the United States and the Soviet Union clearly demonstrates that the world can step back from the brink of destruction. If nations can agree to destroy their intermediate range missiles, they can agree to destroy other weapons as well. They have the power to change the course of history by systematically eliminating all of the world's weapons of mass destruction.

Unfortunately, arms reduction agreements between nations are seriously limited. They depend on the goodwill of the participating parties but they lack a mechanism that will assure compliance. They are not insulated from the changing interpretations that may be proclaimed by a new generation of leaders. They are not part of a comprehensive system designed to provide lasting, global protection from weapons of mass destruction. In some ways, arms control talks are an easy way out for national leaders. By sitting down to negotiate, they can appear to be resolutely striving to eliminate the threat of nuclear war and, at the same time, they can avoid the hard decisions—and the risk—involved in seeking a sound and lasting solution to the problem. The situation has much in common with that of a patient who has been told of a life-threatening cancer. If the cancer is surgically removed, the patient has a very good chance of making a complete recovery and living a normal life. But the operation is difficult and involves an element of risk. The patient is concerned about the operation and begins a frantic search for a risk-free alternative. With luck, the patient will abandon the search and consent to surgery before it is too late. But if the search is prolonged, if the cancer is allowed to spread, death is a certainty.

The cancer of nuclear arms has been spreading for more than four decades and the supposed remedy of arms agreements has been little more than a placebo. Kenneth Adelman, then director of the U. S. Arms Control and Disarmament Agency, was quoted in the September 14, 1987, *Wall Street Journal* as lamenting the way arms control has been treated as the primary source of peace. "The public, the press and the Congress mistake peace for arms control," said Adelman. "The two things have scant to do with each other." Arms reduction negotiations have given the world a false sense of security. They have lulled people into believing that the life or death problem of nuclear weapons can be solved through unenforceable treaties that nations may violate whenever it appears to be in their best interest. Why would any nation—or any individual, for that matter—abide by an agreement when it lacks a penalty clause, when it is void of any system for penalizing violators? If a treaty is violated by one nation, the other nation responds with a violation of its own. For example, the construction by the Soviet Union of radar installations that appear to violate arms treaties has provided a reason for the United States to feel justified about proceeding with violations of its own. Even if negotiations continue to succeed in bringing about a reduction in the number of weapons, the nuclear danger would remain. If nuclear arsenals were reduced by 50 percent, there would still be enough weapons to kill every person on earth five times. If the reduction reached 99 percent, the United States and the Soviet Union each would continue to possess at least 20 thermonuclear devices, each 200 times more powerful than the bomb dropped on Hiroshima.

> **. . . it can be argued that arms control negotiations have contributed to world-wide nuclear tyranny because they have been based on national expediency instead of international law.**

At the 99 percent reduction level—a level not foreseen in the most optimistic appraisals of the arms control process—the world still would be faced with a threat equal to the power of at least 4,000 Hiroshima bombs. For those concerned about the survival of civilization, such a threat ought to be totally unacceptable. Shrink the world's stockpile of nuclear weapons to two—one for each of the superpowers—and there still would be no guarantee that one or both would not be detonated with appalling consequences. Whether there are two or 2,000 nuclear weapons, we have only two choices. We can tolerate the situation and accept continued danger or we can move in new and productive ways to end the threat once and for all. Arms reduction negotiations are not new and they often have been counterproductive. In fact, it can be argued that arms control negotiations have contributed to world-wide nuclear tyranny because they have been based on national expediency instead of international law.

In an article in the Spring, 1985, issue of *World Policy Journal*, Robert C. Johansen noted, "Paradoxically, one reason for the failure of arms control has been its almost exclusive reliance on negotiations as the route to progress—a reliance that has tended to exacerbate rather than diminish superpower distrust." Dr.

Johansen, an internationally recognized authority on disarmament issues, also pointed out that, "If negotiations fail, the public and officials mistakenly assume that nothing else can be done to decrease security threats. In fact, there is much the United States can do on its own without any negotiations at all . . . ." Arms agreement as the world has experienced it and as presently negotiated, agreement without enforceable international law to assure compliance and penalize violations, has about as much substance as the emperor's new suit of clothes. In many ways, it is a dangerous approach, not only because it fosters deception and duplicity but because it distracts our attention. It keeps us from focusing on an ancient and attainable goal, peace based on international law. This goal is closer today than ever before because, as Edna St. Vincent Millay put it years ago, "There are no islands any more." The realities of today's world have put an end to isolation. They have set the stage for peace based on the global extension of the rule of law. Much progress already has been made. In his book, *A Common Sense Guide to World Peace* (Oceana Publications, Inc., New York, New York, 1985), Professor Benjamin B. Ferencz points out many encouraging examples of the strengthening of international law. "The number of nations joining in cooperative international efforts has been consistently increasing," said Professor Ferencz. "Each advance—though inadequate—was an improvement on the past. The line of progress is there for all to see." Professor Ferencz lists such developments as the Law of the Sea treaty, agreements on the exploration and utilization of outer space, cooperation in the battle against hunger and disease through such agencies as the World Health Organization and, interestingly, promotion and control of non-military use of atomic energy through the International Atomic Energy Agency. Nations voluntarily ratified these agreements because they perceived them to be in their own best interests. Their willingness to take such action certainly seems to indicate a growing awareness of the importance of the rule of law in international affairs.

There is, of course, one obvious difference between existing international agreements and the law needed to outlaw weapons of mass destruction. The existing agreements depend on voluntary compliance. The proposed anti-weapons law will assure compliance through rewards and penalties. The proposed anti-weapons law as a part of the Civilized Defense Plan is a logical step in the evolutionary development of an international rule of law. The proposed law deals with the most serious of subjects—the survival of life itself. The law that will preserve the earth and its people cannot be equated with rules for collecting minerals from the ocean floor. The latter can be left to voluntary compliance. The former must include a mechanism for universal enforcement.

In nearly every conceivable way, the nations of earth have moved well beyond the days of imaginary self-sufficiency. In communication and transportation, for example, satellites assist the navigators of air and ocean, broadcast radio and television signals to and from the most distant locations and provide the connections that enable human beings—and the computers human beings have created—to talk across the miles. Technology has knocked down the walls between nations, facilitating an accelerating exchange of information. The flow is so powerful, its potential benefits so great, that it is beyond resistance. It has penetrated the traditional insularity of the Soviet Union and the People's Republic of China, created a climate of expectation in Africa and Latin America and suggested the fulfillment of Wendell Willkie's World War II prophecy of One World. In a world without islands—the "global village" described by Marshall McLuhan— global law is a necessity. Law is the only assurance of justice, the only barrier against violence and oppression. Law offers the only hope of taking the gun out of the hand of the aggressor, ending for all time the threat of war and preserving the earth and its people.

The prize of peace through international agreement in the form of law is so desirable that it is difficult to understand why it repeatedly has slipped through humanity's fingers. The high hopes of the past—from the Hague Conventions to the League of Nations to the United Nations—have not been realized. Treaties that supposedly have the force of law, but were not law, have been signed, sealed and ignored. Aggressive military action continues to be the prerogative of any nation and without an effective mechanism for judging a nation's actions and a means of enforcing the judgment, how could it be otherwise? In a world without an accepted international judicial or enforcement system, no nation will risk its independence on the strength of a piece of paper and a handshake. National leaders will not jeopardize the safety of their fellow citizens—and their own careers—by gambling on the sincerity of their counterparts across the frontier. Asking the nations of the world to base the elimination of weapons of mass destruction on a foundation of mutual trust is as utopian as asking a family to entrust the safety of its home and possessions to the goodwill of a population that inevitably includes a burglar or two. The family quite wisely will prefer to rely on the rule of law, backed by the defense of a sturdy lock and an effective police force.  The family knows that enforceable law provides the framework in which trust can operate—and what is true for a family is equally true for the family of nations.  Once this fundamental fact of international relations is understood, it is apparent what must be done.  Humanity must discover a way to guarantee agreements between sovereign states.  Humanity must create an environment that encourages trust among nations, a system that provides ironclad assurance against aggression.

> **Without surrendering national identities, we must create a program that will lead to the supremacy of world law, a program which must include a powerful, yet humane, system of enforcement.**

The development of such a system surely is not beyond the capacity of the human mind.  The problem is that some of our best minds have been concentrating on the perpetuation of the balance of nuclear terror.  They have been planning and building weapons, theorizing about the desirability of first or second strike strategies, arguing about what is rather than what might be.  They have ignored the fundamental issue of building lasting international security.  Soon after World War II, civic leader Basil O'Connor warned, "The world cannot continue to wage war like physical giants and to seek peace like intellectual pygmies."  It is a warning the world must heed.  We must mobilize our intellectual resources in the cause of peace.  We must break the constraining bonds of the past and begin thinking in terms of the future.  Without surrendering national identities, we must create a program that will lead to the supremacy of world law, a program that must include a powerful, yet humane, system of enforcement.  Such a system must be based on a force that is stronger than the military, a force upon which armed might depends: the force of economic power.  The United States, in cooperation with other nations, possesses vast economic resources that can be utilized in the creation and enforcement of an international law that will effectively eliminate weapons of mass destruction.  The United States also has a tradition of respect for the individual and belief in the rule of law.  These values have contributed to U.S. success.  They also have created the leadership responsibilities that we next will examine.

# 5

# The Question
# Of
# U.S. Leadership

*We have it in our power to begin the world again.*
—Thomas Paine

One of the greatest surprises I experienced during the development of the Civilized Defense Plan was the reaction of many well-nformed, intelligent Americans to the Plan's first premise. These people objected to the idea that the United States has any special responsibility to provide the leadership the world needs. Some of the objections were based on a negative perception of the role the United States has played in international affairs since the end of World War II. Those who hold this view tend to blame the United States for most of the world's problems or, at the very least, to lump together the United States and the Soviet Union as equal sources of a rivalry that has had catastrophic impact on society. Some Americans who objected maintained that the United States has no right to attempt to influence other nations, that it has no monopoly on goodness or wisdom and that it should let fellow members of the international community make up their minds without U.S. inducements or sanctions. Still others objected strongly to any suggestion that U.S. responsibility might be related to the leadership role our country played in the development and use of nuclear weapons. These objectors—and they well may be speaking for a majority of Americans—believe the United States was justified in creating and using nuclear bombs. They reject any notion of guilt or reparation.

I was surprised by the number and scope of the objections because I believed U.S. responsibility was self-evident. I believed—and still believe—that U.S. leadership long has had a profoundly positive influence on world affairs, that it is essential for America to use its influence to solve humanity's most challenging and threatening problem. Of course, the United States has made its share of mistakes. From time to time, it has blundered and fumbled in its relations with other nations. Worst of all, it sometimes has forgotten the principles that guided its founding and undergirded its progress. It seems to me that the United States is simply a very human country. It has all the vices and virtues of the individual human being. It can be meek and arrogant, understanding and inconsiderate, generous and stingy. It can demonstrate both its love for peace and, from time to time, its eagerness to throw its weight around. But as with an individual, a nation should not be judged through some rigorous calculation of its virtues and its vices. It should be judged on the sum total of its influence on society. In the case of the United States, I believe it will come through such a judgment with flying colors. Any objective review of history will show that over the years the United States has set a rather admirable standard of behavior. Americans have been reluctant to become involved in war and, for the most part, they have resisted the temptation to use military power to expand America's territory. At the same time, they have created a society that enjoys the admiration of much of the world. America has become an example of what freedom can accomplish. America is not blameless in its domestic or foreign affairs. But those who question U.S. responsibility to provide leadership ought to ask themselves: *What would the world be like if the United States had never existed?*

Depending on personal perspectives and experiences, each individual undoubtedly will have a different answer. But I am confident that many of these answers will be built around the fundamental themes of freedom and diversity. For more than two centuries, the United States has provided the world with a lesson in what can be accomplished through a commitment to the freedom of the individual. This commitment has fostered healthy competition, stimulated social and economic initiatives and built a society that, for all of its imperfections, remains the idol of most of the rest of the world. The American commitment to freedom has attracted the homeless, the hungry and the oppressed of other countries. It has lured to our shores innovators and idealists, artists and intellectuals. It has been a magnet to men and women of many cultures, to those who share a desire to make up their own minds about what is in their best interest. This commitment to freedom has contributed to U.S. economic strength, created a voluntary sector that is as important as it is unique and encouraged the establishment of religious and educational institutions that satisfy the widest range of spiritual and educational needs.

All of these accomplishments of a society dedicated to freedom have been watched and emulated by most of the rest of the world. And when the world's own freedom was in jeopardy, it knew it could count on the United States. Its citizens fought wars to preserve the freedom of others and, in the post-World War II period, American taxpayers built and maintained a powerful defense machine to protect freedom around the world. What would the world be like if the United States had never existed? Would the common language of Western Europe be German or Russian? Would the world's dominant political philosophy be fascism or communism? Would Japan be an economic superpower as it is today or the expansionist military state that it once aspired to be? Would any of the world's nations be able to provide humanity with a functioning example of freedom and what it can accomplish?

And what about the example of productive diversity that has been provided by the United States? Americans have demonstrated that people of differing backgrounds, beliefs and attainments can live together in a lasting and mutually beneficial relationship. They have shown that diversity can enrich a nation, that ethnic, cultural, religious and other differences can serve to strengthen a society—if that society understands both the importance of the individual and the supremacy of law.

In the world society that is emerging, the individual will play a key role. The individual will be the catalyst that will enable institutions to adapt to accelerating change. The individual will provide the leadership that will move governments away from the dangerous status quo and bring them into harmony with the reality of a world that is being transformed by the relentless impact of such forces as technology and trade. But the individual leaders of today and tomorrow need a continuing example. The world's leaders need the leadership of the United States.

No project on which I have worked has succeeded without leadership. It is the one essential ingredient. Without someone to establish goals and provide direction, confusion and failure are inevitable. The dynamics of leadership are among society's mysteries. But while we do not fully understand leadership, we do know many of its hallmarks. True leaders assume responsibility. They believe in themselves and in something higher than themselves. They strive to help others. They think positively about their cause. They are convinced they can make the future better than the present or the past. In our world society, the United States continues to occupy the leadership position. It is the only economic and military superpower. In every aspect of international affairs, it can deal from a position of strength. The United States, with its tradition of freedom and diversity under law, can lead the way into the new world of the Fourth Freedom—if only the American people have the will to utilize their own enormous ability and employ it for the benefit of the nations that look to the United States for leadership. I believe there is a tremendous world-wide reservoir of admiration for the United States and what it stands for. Of course, the admirers get far less attention than the demonstrators who burn our flag and shout, "Yankee go home." But from my experience talking to business people and everyday citizens in countries around the globe, I can tell you that America's supporters outnumber its detractors.

> **In our world society, the United States continues to occupy the leadership position. It is the only economic and military power. With its tradition of freedom and diversity under law, it can lead the way into the new world of the Fourth Freedom.**

Through the years, I have witnessed countless examples of the support that exists for America and its initiatives. Perhaps the most memorable incident occurred in 1960 when my wife and I spent a week vacationing at Bergenstock, high above Lake Lucerne in Switzerland. Most of the resort's guests were French, German or Italian and, on one strikingly beautiful night, nearly all of us were out dancing under the stars. All at once, the music stopped. High above us, we could see a bright object moving toward us across the heavens. It was going to pass directly over our heads and everyone seemed to know that it was Echo, the reflecting satellite that the United States had put into orbit. Where only minutes before the night was brightened by music, dancing and conversation, there now was silence. Then as Echo streaked over us, there was a spontaneous salute. Those who had been sitting at tables rose as one person to join the dancers, raising their glasses in a toast we could understand: "Long live the United States of America!" Never have I been more proud to be an American.

The Western Europeans we were with that night knew the Soviet Union had launched Sputnik, the first object to achieve earth orbit. But they were cheering for the United States to catch up. They were applauding our attempt to assume leadership, and the applause grew louder in 1969 when we reasserted our leadership in space exploration by a lunar landing. I am convinced the world's applause is as real today as it was two decades ago. I have heard it in the words and actions of individuals from many countries and through a range of business and social relationships. I also have heard criticism, especially from our European friends. They understand that Americans are deeply imprinted with the European cultural tradition. The United States is Europe's child and, like all parents, the Europeans sometimes are quick to criticize. They expect their child to fulfill many of their hopes. They believe America is destined to excel, to lead the world into a new age of achievement. If Americans prudently utilize their enormous economic power, if they strive for conciliation and spurn the temptation to say "I told you so" to those nations impoverished by the failed, freedom-stifling policies of totalitarian regimes, American leadership will be increasingly accepted in Europe and in other regions as well. If the United States renews its commitment to the values that contributed to its greatness, it can become instrumental in building the international cooperation that will bring to all nations the priceless gift of lasting security.

# PREMISE II

## Using Economic Power
## To Achieve
## Lasting International Security

As the keystone of today's interconnected and interdependent world economy, the United States has the influence to enlist the support of other nations in the development of a world law that would eliminate weapons of mass destruction through economic incentives and, if necessary, the imposition of sanctions on trade in manufactured goods. To insure the effectiveness of these sanctions and in sharp contrast with the sanctions of the past, they would apply uniformly to nations unwilling to support the law *and to nations that continue to trade with nations not supporting the law.*

# 6

# Building Political Muscle Out Of Economic Strength

*You only get political answers if you go through the economic answers first.*

— Lord Cockfield

A recent phenomenon—one that I suspect all of us find difficult to ignore—is the frequent pilgrimages of U.S. governors and mayors to Japan. They make these trips with great fanfare and without the slightest embarrassment about the hat-in-hand image that they project. In fact, they seem to take pride in playing the beggar's role. They want their constituents to know that they are seeking to attract Japanese investment and, when election day rolls around, they hope the voters will remember this shrewd effort to bolster the local economy. Voters, like governors, mayors and all the rest of us, know that Japanese industry has become a significant factor in the American marketplace, a source of jobs as well as goods. The economic growth of Japan, its rapid progress toward world economic leadership, is acknowledged in America and around the globe. It also is important for us to recognize that the rise of Japan is no accident. It is a development directly related to the underlying strategy of the Civilized Defense Plan.

In the 1930s and '40s, Japan sought to solve its problems through armed aggression. With an expanding population and limited natural resources, it undertook the conquest of the Asian mainland and the island territories of the Pacific's western rim.

41

After an exhausting effort that consumed incredible quantities of Japan's human and material resources, the effort failed. Only the problems remained. With the option of a military solution no longer available, thanks to a constitutional limitation on arms expenditures, Japan sought to solve its problems by peaceful means, by launching an economic offensive on the markets of the world. As all of us are well aware, the results have been extraordinary.

Authorities on world commerce cite a number of factors in explaining the rapid rise of Japan in relation to the United States. Many of these factors have their origin in Japan's modest expenditures for armaments. The Japanese spend somewhere in the neighborhood of one percent of their GNP on defense while, as I noted earlier, Americans spend approximately seven percent and the Soviets a considerably higher percentage. As a result, Japan can invest its wealth in the continuous modernization of its industrial base, the research and development that leads to desirable new civilian products and the refinement of aggressive worldwide trading policies and practices. Meanwhile, the superpowers continue to squander their wealth on weapons they vow never to use. While the economies of the superpowers stagnate, as in the Soviet Union, or struggle to achieve gains, as in the United States, both nations continue to pour their resources into underground silos and reinforced bunkers. They continue to give budgetary priority to military installations valued for their destructive, rather than productive, potential.

If constantly increasing expenditures for defense resulted in true national security, they could be justified. But a glance at today's world indicates that a sense of security has eluded both the United States and the Soviet Union. How can there be security for either nation when their economies are showing signs of distress? How can the military be strong when the bedrock of the economy may be starting to fracture?

The U. S. economy, of course, is in far better shape than the economy of the Soviet Union. Despite the problems of recent years—the substantial federal debt and the unprecedented trade deficit—the United States remains the number one economic power. Even the federal debt and trade deficit problems lose some of their frightfulness when we view them from an analytical perspective. In an article in the March, 1989, *Reader's Digest*, Nobel laureate economist Milton Friedman pointed out, "As a percentage of the national income, the deficit is not out of line with levels frequently reached in the past. Indeed, if the surplus at state and local levels is subtracted, the combined government deficit is considerably lower than in many past years." Dr. Friedman also has some reassuring views on the foreign trade deficit. He notes that ". . . prospects for domestic investment have been sufficiently bright to make investment in the United States more attractive—to American banks and to foreigners—than investment elsewhere . . . . It is a mystery to me why, to take a specific example, it is regarded as a sign of Japanese strength and U.S. weakness that the Japanese find it attractive to invest so much in the United States. Surely, it is precisely the reverse—a sign of U.S. strength and Japanese weakness."

Indeed, all generally accepted measurements show that the U.S. economy is roughly twice the size of Japan's and the United States is number one in both imports and exports. During the past few years, the United States has created more new jobs than Japan and Western Europe combined. This growth in employment helps account for the much-discussed trade imbalance. More Americans have jobs than ever before and they are using their paychecks to buy more things, including the products of other nations. This growing ability to buy contrasts sharply with the situation abroad. In many other nations, including Japan, economic growth has been minimal in recent years, strengthening the position of the United States as the world's most desirable market. The appetite of Americans for such products as automobiles, telephones, televisions, and other electronic equipment, light bulbs, office machines, furniture, and clothing is so strong that it has attracted the attention of marketeers from around the world. The nations of the European Community, the Soviet bloc, the Far East, Africa, Latin America—indeed, nations in every corner of the globe—are eager to compete for a share of our purchases.

On the production side of the ledger, the agricultural segment of the American economy ranks first in the export of corn, soybeans, wheat, cotton, and other products. American industry leads in exporting airplanes, computers, data-processing equipment and a variety of other types of machinery and equipment, including, it should be noted, nuclear reactors. As a business executive, I know that most manufacturing sectors of the U.S. economy are doing well in holding their own against competitors. I have observed this fact in the marketplace and I am not surprised that my observation has been confirmed by objective authorities. For example, Britain's National Institute of Economic and Social Research has reported that U.S. labor costs are now the most competitive in the world. U.S. manufacturing productivity has been rising steadily throughout the 1980s and, to assure a continuing rise, U.S. business is increasing its spending on new plants and equipment. Unfortunately, much of the business news transmitted to the American people paints a far less upbeat picture. More attention is given to U.S. economic problems—real and imagined—than to its impressive economic achievements. Business reporters seem fascinated by an October stock market plunge and bored by month after month of overall economic growth. As we should have learned to expect, bad news—whether about the economy or any other subject—tends to drive good news out of the news-papers and newscasts. Ironically, the mass media that often has failed to give Americans an accurate account of the strength of their economy has been the messenger of America's richness to the citizens of other lands. Jack Valenti of the Motion Picture Export Association has noted that even films that present a less-than-attractive view of American society can have a positive impact. He cites as an example the vintage film, *The Grapes of Wrath*. While Americans were distressed by the film's painful portrayal of poverty in their country, many non-Americans were impressed to learn from the film that in America even the poorest individuals own automobiles.

The power of the American example has been especially evi-dent in the development of the dynamic economies of many of the nations of the western Pacific. They have borrowed heavily from American business concepts and they have benefited from the strength of American markets. These countries, like almost every other corner of the earth, also have been much influenced by American popular culture—a culture that reflects the inno-vation that is the hallmark of American entrepreneurship. McDonald's golden arches have turned up in Paris, Rome, Nairobi, Tokyo, and Moscow—in 42 nations in nearly every corner of the world. *What They Don't Teach You at the Harvard Business School* was a bestseller in Bangkok and the Super Bowl was wit-nessed by viewers in 28 countries from Australia to Venezuela. Even in the Soviet Union, American influence is apparent. Levi's jeans are a hot fashion item and—much to the mystification of people of my generation—American popular music is heard, enjoyed and imitated.

Perhaps the most significant evidence of the power of the American example has been the move toward limited private economic initiative on the part of the Soviet Union and the People's Republic of China—a move that well may have been influenced by the stories of success written on the American pattern by such nations as South Korea, Hong Kong and Singapore. Without abandoning their public commitment to Marxist-Leninist economic theory, China and the Soviet Union have appeared to be edging toward acceptance of at least a few elements of a market-oriented individual incentive system. By beginning to tolerate a modest amount of private enterprise, the leaders of these two communist nations were acknowledging economic shortcomings and the world will be watching to see if the movement toward a market economy is maintained, accelerated or reversed.

For the Soviet Union, a nation saddled with the cost of a continually expanding arsenal of nuclear and other civilian-threatening weapons, the problems are especially obvious. Industrial production is stagnant and the agricultural segment of the economy has had enormous difficulty in meeting the nutritional requirements of Soviet families. Indeed, some authorities believe Soviet grain production may be lower now than it was a decade ago. Soviet-made consumer goods are notoriously shoddy. Only in armaments production and the exploration of outer space can the Soviets make any claim to world leadership.

> **By shifting the focus of international rivalry from armed confrontation to economic competition, the United States can help create a world in which every nation can help make its own non-violent claim to success.**

On the economic front, the United States has a clear advantage. It possesses power which can influence the support of friends and, at the very least, gain the undivided attention of enemies. Unlike the weapons of the military, devices that produce only death and destruction, economic power can enhance life and foster progress. By shifting the focus of international rivalry from armed confrontation to economic competition, the United States can help create a world in which every nation stands to prosper. Competition under the rule of law—or the rules of the game—whether in business or sports, usually brings out the best in all the participants. Free of the burden of expenditures for weapons of mass destruction, every nation would be able to do its best to achieve its full potential, to make its own non-violent claim to success. In a world without catastrophic aggressive weapons, every nation would continue to have the right to defend itself and the defensive abilities of all nations would be enhanced. In a world without weapons of mass destruction, it is questionable if many nations would be eager to continue some of the joint defensive arrangements that were products of East-West tensions. For example, with the understanding that the United States would continue to be able to sell them the defensive weapons they deem appropriate, would not Japan, West Germany and South Korea be ready to assume responsibility for their own defense?

Given the disparity between the economies of the United States and the Soviet Union, it is tragic that the United States has failed to take advantage of its opportunities. Instead of playing its strong suit—an economy that has achieved so much that it is increasingly imitated around the world—the United States has played to Soviet strength by joining in an arms race in which both nations stand to lose. It is, in my opinion, utterly absurd that the greatest proponent of Marxist-Leninist socialism and the greatest proponent of Adam Smith's capitalism are competing in the field of arms rather than in the field of economics.

The United States has been magnificently successful in exporting both its products and its ideas. It has played a key role in creating a world in which trade has become an increasingly essential element. Yet, it has not understood the necessity of utilizing trade in the cause of international security. It has not exerted its leadership by shifting its focus from a defense based on military power to a defense that utilizes the power of the American economy. By emphasizing economic rather than military power, the United States can enhance its ideological initiative. By speaking up for the powerless who have been victims or potential victims of the machines of aggression, the United States could reassert the idealism that long was an admired hallmark of American policy. The nation that always has professed to be the champion of freedom should make clear its unshakable commitment to the Fourth Freedom. But the time for this bold humanitarian action is limited. The window of opportunity will not stay open forever. Each day wasted on the disproved policies of the past moves the world closer to catastrophe.

It is ironic that the nation that delights in innovation has been far from innovative in its defense of freedom, including Freedom from Fear. The nation that developed incredible consistency in its hamburgers and French fries has been inconsistent in the development of policies that will end the threat of global disaster. The nation that leads in per capita college attendance, in the number of Nobel laureates and in opening its doors to the world's oppressed, has failed to lead in finding a way of protecting the people of all nations from the greatest threats to their existence—threats that go beyond the loss of life because they endanger even the hope of posterity. Fortunately for humanity, the failure of the United States need not be permanent. It can still cooperate with other nations in leading the world to a place of common safety, to an age that is free of the weapons of mass destruction and vaccinated against the disease of terrorism. But we must take a new road, one that follows a direction determined by the economic interdependence of nations. It is a road built on enforcement of international law by offering incentives to law abiding nations and by outlawing trade with outlaw nations. This is the road we will explore in the pages that follow.

# 7

# Backing Words
# With Actions

*Action may not always bring happiness; but there is no
happiness without action.*

—Benjamin Disraeli

As a business manager, I long ago learned that one of the
greatest challenges is to achieve a balance between words and
actions, to understand that they are the opposite sides of the same
coin. Together they are a medium of exchange, a power for getting things done. Separately, their power is limited. For example,
I have noticed that a problem can become very difficult to solve
if its description is too wordy. Under the influence of superfluous
words, a specific problem becomes increasingly complex and soon
it appears to be linked to an array of other problems, real or perceived. The pile of problems or potential problems seems insurmountable and the future of a promising project suddenly is in
jeopardy. If you take the time to analyze the situation, you often
will find that minor or even non-existent obstacles are given the
same weight as obstacles of major importance. You also will find
that the major obstacles are few in number and that they rarely
are overcome by those who spend the most time talking about
them. I think many business managers would agree with me that
those who talk the most tend to accomplish the least, while those
who talk the least tend to accomplish the most. More often than
not, the talkers seem to have a fixation on the negative. My
experience tells me that negative thinking is more prevalent than

positive thinking and I am convinced that if someone wrote a book entitled *The Power of Negative Thinking*, it would be an overnight best seller. The buyers surely would include all the people whose negative thoughts keep them from doing what they want to do or being what they want to be.

In international relations, negative thinking has played a role in shaping conventional views of the potential effectiveness of economic sanctions. Many academicians, foreign policy analysts, business executives and others seem incapable of understanding that the failed sanctions of the past were strong on words and weak on action. Many of these people—and I have talked to dozens of them—refuse to visualize what might be. They insist on identifying obstacles, or imagined obstacles, to the successful implementation of sanctions. The objections almost always are built around what I have come to describe as the "what if?" question:

"What if Japan won't join us in sanctioning international lawbreakers."

"What if France decides to go it alone?"

"What if any of our other trading partners in Europe or on the Pacific Rim balk at sanctioning nations unwilling to support laws outlawing weapons of mass destruction and prohibiting the exportation of terrorism?"

The best answer to these and other what-ifs is another question: Why would?

Why would Japan give up access to its best market and, at the same time, reject the umbrella of protection extended by the United States?

Why would France, devastated by two world wars, turn its back on a new opportunity to achieve lasting security?

Why would any trading partner of the United States reject the opportunity to take part in the creation of international law that would protect them from civilian-threatening weapons, end aggression, prohibit terrorism and foster continued economic development? Why would they turn down a proposal that only requires them to withhold trade in manufactured goods from nations whose leaders are unwilling to provide their own people with the protection afforded by international law? For that matter, why would the Soviet Union not want to support the proposed anti-weapons law? Why would Soviet leaders not leap at the chance to lay down some of the heavy burden of armaments and enjoy the benefits of enhanced international trade?

Now I am sure those who ask the "what if?" questions are in good faith. But in their effort to raise objections, they tend to overlook the positive aspects of the economic sanctions called for in the Plan.  As I have said repeatedly, economic sanctions, used prudently and without malice, can be effective in human relationships at every level, from the interpersonal to the international. Sanctions give support to words, often accomplishing what cannot be accomplished by words alone.  Sanctions are a tool that can be used by any person or any organization that seeks to improve conditions that affect them.  Sanctions can be mutually beneficial. They can help those who impose them and those against whom they are imposed.  They contribute to social order by conferring nonviolent power on every individual and every organization.

> **Sanctions are a tool that can be used by any person or any organization that seeks to improve conditions that affect them.**

Sanctions—really another name for the universal act of withholding—have enormous potential in the field of international affairs.  But as the "what if?" questions indicate, skepticism sometimes makes it difficult for us to see that this potential can be realized in our own age of international interdependence.  Of course, there is nothing new about the concept of economic sanctions.  They are as old as disputes between tribes, cities and kingdoms.  The siege is a form of economic sanction.  So is the blockade.  Both have been used effectively over the years.  The boycott is another example of an economic sanction.  By declining to purchase the goods produced by a target company, individuals can make their voices heard.  They can change company policy—for example, force a multinational corporation to stop promoting the sale of infant formula in developing countries or coerce a U. S. manufacturer into giving up its operations in South Africa.

The federal government effectively uses the threat of economic sanctions to enforce laws and regulations.  If a state fails to build a highway to federal specifications, the U. S. Department of Transportation will withhold funds until the deficiencies are corrected.  If a college or university fails to meet the standards established in civil rights or other legislation, the U. S. Department of Education will threaten to cut off the flow of federal dollars.  Another example of an effective economic sanction is the strike.  By withholding services, labor can achieve gains that cannot be achieved at the bargaining table.  The withholding of goods or services, whether by an individual or by a group, is the most frequently used and most productive strategy for obtaining desired results. As I pointed out earlier in this book, the power to withhold is used day after day by all of us in many of our relations with others.

Withholding works because it is based on the existence of something desirable. This desirable something is the other, positive side of withholding, the incentive that encourages cooperation, the benefit that attracts support. In family relationships, the negative act of withholding draws its power from the desirability of positive acts of affection and support. In industry, a strike can work if both workers and management perceive the mutual benefit of continued efficient production. In international relationships, effective sanctions depend not so much on coercion as on an understanding of the rewards that belong to all nations that maintain their good standing in the world-wide trading community.

Selective employment of the power of withholding—at least the negative side of this power—seems to enjoy wide acceptance in international affairs. One example is the current campaign to impose sweeping economic sanctions on South Africa. Those who support these sanctions believe they will force the South African government to change its racial policies. Supporters of sanctions against South Africa have no doubts about the effectiveness of withholding goods and services. But there is no indication that they have thought about the possibility of using the power of withholding on a world-wide basis, of employing it to eliminate weapons of mass destruction, state-sponsored terrorism and other aggressive behavior—causes that clearly are in the best interests of all nations.

---

**If economic sanctions have a bad name—and they do among some political scientists and historians—the blame must be assigned not to the concept of sanctions but to those responsible for transforming the concept into action.**

---

If economic sanctions have a bad name—and they do among some political scientists and historians—the blame must be assigned not to the concept of sanctions but to those responsibile for transforming the concept into action. To bring this point into focus, we can quickly review the history of the decades after the first World War. During the 1920s and '30s, the use of sanctions was authorized by the Covenant of the League of Nations. League members were expected to impose sanctions on any nation that disturbed the peace. They were supposed to establish a blockade of the aggressor nation's frontiers and to boycott the aggressor's products. To minimize the hardship on nations imposing sanctions, League members were expected to help one another. With prompt, coordinated action by nations professing League membership, economic power might have put an end to war long before the Manhattan Project released the nuclear terror.

In 1931, when Japan invaded the Chinese province of Manchuria, League members scolded Japan and condemned its aggression. They did little else. So Japan consolidated its conquest and, in a gesture of contempt for nations too timid to back their words with actions, withdrew from the League. In 1935, Mussolini's Italy invaded Ethiopia. This act of aggression was an unmistakable violation of the League Covenant and there soon was a flurry of activity. For a short time, it appeared that there would be prompt, coordinated action, that economic sanctions would be imposed against Italy and assistance would be directed to Ethiopia. Public opinion was outspoken in its condemnation of the Italian attack on a state that was weak and nearly defenseless. But public opinion failed to win the support of the governments that could have made sanctions effective. Both France and Britain dragged their feet. The United States, which had declined to become a member of the League, banned shipments of arms to both Italy and Ethiopia. But trade in other goods was allowed to continue. In fact, as Robin Renwick points out in *Economic Sanctions*, a report published in 1985 by the Center for International Affairs, "The United States, which had supplied 6.5 percent of Italy's oil, was supplying 17.8 percent by the end of the year." Without the support of France and Britain—and with the United States providing Italy with an increasing share of the oil needed by its invading army—the League's sanctions lacked decisive impact and the effort soon was abandoned by smaller countries.

When the United Nations Organization was created, there was new talk of using economic measures to keep the peace. A provision for sanctions was built into the U. N. charter. However, implementation demanded the unanimous consent of the Security Council, a body that soon found itself paralyzed by vetoes. As long as the veto power remains intact, there is no possibility of the United Nations imposing sanctions against any major power no matter how barbarous its behavior. The veto, a concession to an outmoded understanding of national interests, has effectively

short-circuited much of the peace-keeping power of the United
Nations. The Security Council has become a theater for futile
East-West debate instead of a builder and maintainer of world
peace.

Again and again, multinational efforts to utilize sanctions have
been aborted by a failure of will. The leaders who undermined
the League of Nations' sanctions and later agreed to the paralyzing Security Council veto demonstrated a remarkable combination
of cowardice and lack of vision. They have not backed their
words with action. They have helped keep the world from seeing
that economic power is the only force that can keep military
power at bay. Even with the inconsistent, fragmented and often
inept economic sanction attempts of the post World War I period,
the record is not as bleak as some observers would have you
believe. In a pioneering comprehensive analysis of the sanctions
issue, *Economic Sanctions in Support of Foreign Policy Goals*
(Institute for International Economics, Washington, D.C., 1983),
Gary Clyde Hufbauer and Jeffery J. Schott took a close look at 78
examples from 1914 to 1983 and concluded "Perhaps surprisingly,
sanctions have been successful in 40 percent of the cases." The
authors, who based their conclusions on a careful analysis of goals
and results, offer a number of explanations for the success or
failure of an attempt to impose sanctions. While some of these
explanations may be debatable, it is difficult to ignore the factual
evidence presented in the Hufbauer-Schott study. Contrary to
popular belief, international sanctions that are properly conceived
and painstakingly implemented have worked in the past and, with
the cooperation of key members of the world community, they can
work in the future.

However, in today's interconnected, global economy, sanctions
rarely can be effective if they are imposed unilaterally. To produce the desired results, sanctions must be multinational and they
must be directed at both the nation that violates an accepted
standard of behavior and its accomplices in crime. Unilateral
sanctions directed at a violator but not at those continuing to
trade with the violator almost always will fail to achieve their
objectives. When Nation A withholds products from Nation B,
Nations C and D too often seem ready to provide a substitute
source for whatever has been withheld. In this regard, the experience of the United States is instructive. Its efforts to impose
sanctions repeatedly have been frustrated by a lack of cooperation
on the part of its allies. When the United States sought to punish
the Soviet Union for its invasion of Afghanistan by withholding

shipments of wheat, the principal losers were American farmers while the Soviets purchased adequate supplies from other countries. The allies are not entirely to blame, either. Without the solid support of enforceable international law, it is tempting to vacillate rather than act, to ignore acts of aggression, to wish away threats to international security. In addition, U. S. policy rarely has been consistent enough to inspire the confidence of friends. In its relations with the Soviets, the United States has moved from outspoken hostility to detente to the present almost undefinable situation. And while U. S. leaders talk about the importance of world-wide economic revitalization, about the triumph of the free market ideas of Adam Smith, the United States continues to give the highest priority to competition in the field of weaponry. It was this competition for weapons superiority that led to the imposition by the United States and its allies of stringent controls on the transfer of advanced technology to the Soviet bloc. While these controls do not seem to have greatly impeded the continued development of Soviet military power, they well may have played an important role in preventing the overall Soviet economy from keeping pace with many other nations. Without access to high speed computers, sophisticated software and other tools of modern industry, the Soviets and their allies have been unable to achieve the economic growth needed to satisfy the needs of their citizens, a development that is having a profound effect on the internal and external affairs of the Soviet Union and nations in its orbit. The West's restrictions on the Eastward flow of technology may prove to be one of history's most successful exercises in the use of sanctions. It may have played an essential role in creating a climate that fosters East-West cooperation.

> **For the first time in human history, economic sanctions against international lawbreakers and their accessories in crime can be fully effective.**

If only we could see the window of opportunity created by this climate and by growing international awareness and economic interdependence. Today the interests of all nations are intertwined as never before. No nation can afford to be cut off from trade with other nations. No nation long can withstand an embargo on shipments of the manufactured goods it needs to meet the needs and aspirations of its people. In simpler times, it might have been possible for a nation to come close to self-sufficiency. But that age now is long past.  People in all parts of the world desire access to the fruits of technology that they see so dramatically revealed on their television screens.  They seek the higher standard of living that is a product of technological advancement. They will not perpetually tolerate a government that prevents them from enjoying what they view as "the good life."  For the first time in human history, economic sanctions against international lawbreakers and their accessories in crime can be fully effective.  Because of the changes wrought by global communications and global trade, the United States can work with other nations to use economic power for the good of all nations. Through judicious use of this power as a persuasive force, the United States and all nations joining with it can achieve genuine national security and move the world into the civilized age of the Fourth Freedom.  To be successful in this most worthwhile endeavor, the law-abiding nations of the world will need to keep their nerve, remembering that they are dealing from a position of strength, that their goal is neither political nor material and that they are acting solely in the service of this generation and those that follow.

The alternative for the United States and its friends is to continue the quest for security by striving to achieve the unattainable goal of military dominance, by blindly following the Soviet lead into permanent second class economic status and by passively accepting the continued threat of terrorism.  Such a course of action surely is unacceptable to Americans and all others who care about themselves, their tradition of freedom and the future of their society.  Those who care constitute a majority in America and many other nations and, I sincerely believe, when they understand the options that are available to them, they will choose development over destruction, competition over conquest, life over death.

# 8

# The Civilizing,
# Security-Building
# Power Of Trade

*We cannot expect to dispose of armaments until we have*
*a plan for common safety.*
— Norman Cousins

In late January or early February of every year, Atlanta plays host to an extraordinary gathering. Those who attend come from nearly every corner of the world, from some 70 nations that, taken together, account for a high percentage of the world's total population and productive resources. Represented are long-time friends and traditional enemies. Delegates come from China and the Soviet Union, India and Pakistan, members of NATO and the Warsaw Pact, Iran, Israel, and the Arab states. Also on hand are representatives from nations of the European Community; Japan, Korea, Taiwan and others on the Pacific Rim; many of the nations of Africa, South and Central America as well as Australia, New Zealand, Mexico, Canada and, of course, the United States. All gather in one place at one time to discuss one subject: How to do a better job of satisfying the growing world-wide demand for chicken and other poultry products.

Those who attend the Southeastern International Poultry Show—and I am one of them—expect to learn the latest developments in the poultry industry. They find out what is happening in the research centers of the world. They exchange ideas with poultry raisers, equipment manufacturers, feed processors, nutritionists, specialists in genetics and all the other disciplines that play a

57

role in poultry production. Through formal and informal meetings, attendees establish networks through which information travels around the globe throughout the year. Like the international expositions of other industries, the Southeastern Show is an economic phenomenon. Those who attend are primarily motivated by economic considerations. Socialists and capitalists alike seek to gather knowledge that will contribute to economic improvement. They aim to increase production efficiency, to produce more pounds of poultry products from fewer pounds of feed with the lowest possible costs for labor, plant and equipment. For those who attend, the subject is exciting. Efficient poultry production is an important goal for many governments. Dwayne Andreas, chairman of the board of a major multi-national agribusiness, has been quoted as saying that when the conversation turns to chicken raising, his friend, Mikhail Gorbachev, sits on the edge of his chair. The concentration on economic gain, useful in itself, also has many positive social consequences. The world's supply of high quality edible protein is increased. The diets of families in developing countries are improved. The cost of poultry products is kept within reach of moderate income families in industrialized and preindustrialized nations. The drudgery of outmoded poultry production methods is lessened. Not least of all, lasting personal relationships are established among poultry people, relationships which transcend political, ideological and sectarian boundaries and which tend to increase international understanding.

> **As Ralph Waldo Emerson observed, "Trade, as all men know, is the antagonist of war."**

Sometimes these relationships ripen into friendships that must, over the long haul, contribute to a reduction in tensions between nations. Indeed, the very thought of friends going to war against friends is reprehensible. International trade, as exemplified by the industry show and similar activities, always has been a civilizing force. Trade penetrates frontiers with desired goods and attractive new ideas. Trade creates markets and stimulates economic competition. Trade opens doors that can give all of earth's people access to all of earth's bounty. As Ralph Waldo Emerson observed, "Trade, as all men know, is the antagonist of war. Trade brings men to look at each other in the face and gives the parties the knowledge that these enemies over sea or over the mountain are such men as we who laugh and grieve, who love and fear as we do."

Trade is a major contributor to the wealth of nations. A nation makes itself and its trading partners richer when it exchanges what it has in abundance for the products it needs. A nation which denies the necessity of trade and isolates itself from the international marketplace inevitably impoverishes its people. China is one of many examples. In striving to achieve self-sufficiency during the post-World War II period, the Chinese economy fell far behind that of other nations, including many with far fewer natural or human resources. Now China is trying to catch up and trade is receiving new emphasis. From my own business contacts with the Chinese, I have come to believe that they increasingly are aware of the evolutionary development of a global economy. They see what has been happening in such places as Taiwan, South Korea and Hong Kong—states that 30 years ago seemed to be among the least likely sites for accelerating industrial development—and the Chinese see no reason why they should continue to take a back seat. To use a somewhat overworked expression, they want a piece of the action. While the tragic events of May and June, 1989, show that Chinese leaders are far from ready to accept even a hint of political change, their actions indicate a desire to play at least some role in a global marketplace that has linked the interests of humanity as never before.

Another case in point is the Soviet Union. For most of the years from the Bolshevik Revolution of 1917 to the present, the Soviets have been relatively minor players on the field of international commerce. They have emphasized trade within their own borders or among the member states of the Council for Mutual Economic Assistance (COMECON). As at least a partial result of this policy, the Soviet Union and other COMECON members have largely been cut off from trade with some of the world's most advanced economies. For example, trade between the European Community and its Eastern European neighbors now accounts for a mere seven percent of the Community's trade and the percentage appears to be declining. There are many reasons for the relatively low level of trade between East and West. One involves the foreign exchange problems of the Soviet Union and other COMECON members. Another is the questionable quality of many of the East's export products. Finally, there is the economic paralysis that seems to be an unavoidable consequence of a rigid and repressive system of economic planning.

Now please understand, I am not indulging in any kind of Soviet bashing nor am I attempting to denigrate any political, economic or ideological point of view. I am trying to be objective. I am outlining the possibilities for tomorrow by describing the reality of yesterday and today.

The repressive nature of the Soviet system of economic planning—indeed, the repression that is evident in many aspects of the Soviet Union—has been pointed out by no less an authority than President Gorbachev. In his push for reform, he in effect has been asking, "If we had a less repressive system, would our economy and our standard of living be so far behind those of our neighbors in Western Europe?" The Soviet people are no different than the people of the United States or Western Europe in their desire for adequate food, shelter and other necessities or in their aspirations to enjoy the blessings of God's creation and the fruits of humanity's imagination. Through the years, I have talked to many Soviet citizens. I am convinced they have the ability to take advantage of the opportunities that inevitably will be created when economic power supplants military power in the strategic thinking of the world's leaders. If the Soviet people are given a chance to do more thinking and acting for themselves—and this certainly seems to be the thrust of the Gorbachev reforms—they will benefit and so will the rest of the world. Without totally abandoning the tenets of socialism, they will be able to play a greater role in the global marketplace. They will be contributors to and recipients of all the benefits of the civilizing power of trade.

In many parts of the world, the reality of the global marketplace is helping induce nations to at least begin to readjust their priorities. They are giving serious consideration to a possible scaling down of their investment in the military in order to give more attention to the development of the domestic economy. Vietnam, for example, now seems to recognize that its military adventures may have been counterproductive. Its attempt to dominate its neighbors through force of arms undermined an already shaky economy, meanwhile, such next door neighbors as Thailand and Singapore were enjoying strong economic growth. That lesson ought to be clear to all national leaders who sincerely seek to improve the lot of their citizens. The military road leads to disaster while the economic road leads to increased opportunity, to a better life for all. Once this fact is understood, it should not be difficult to understand how economic power can be mobilized in the cause of international security. Through the incentives and sanctions called for in CDP, the nations of the world can lift from humanity the terrible threat of weapons that can eliminate life itself.

Let me again emphasize the timeliness of CDP. The world's economy never has been as integrated as it is today. Never before has the economy of a nation been so dependent on the economies of other nations. Western Europe, the United States and Japan already have become, as economic expert Kenichi Ohmai put it in the Spring, 1988, issue of Best of Business magazine ". . . one market, one production zone . . . ." Other nations are well on their way to becoming part of this unified global economic entity. They stand to benefit from the opportunities created by economic globalization. They need protection from civilian-threatening weapons and terrorism. They would be severely affected if they were cut off from trade in manufactured goods. These factors—increased access to the global marketplace and freedom from fear of aggression by force of arms or state-sponsored terrorism—are persuasive positive reasons for the acceptance of CDP by the nations of the world. There also is a negative reason—the possible withholding by the United States and other nations of trade in manufactured goods.

Few nations long could tolerate the withholding of the manufactured products needed by every segment of society. Few nations come close to self-sufficiency, especially in manufacturing. A boycott would be devastating. It would affect imports as well as exports and it would apply both to a nation and its trading partners. So if the United States in concert with other nations found it necessary to impose sanctions against a nation because it refused to support the international anti-weapons law, those sanctions would apply both to the shipment of manufactured goods to that nation and the acceptance of manufactured goods from that nation. The United States and other major markets no longer would be open for business with a nation that refused to support a law designed to achieve true international security.

But what about the United States? I frequently am asked that question by Americans who wonder if their country's economy could survive the refusal by one of its major trading partners to join in supporting the international anti-weapons law. The partners mentioned in the question range from the nations of the European Community to Japan, South Korea, Taiwan, Hong Kong, Singapore, and Thailand. The best response to this what-if question is another question: Why would any nation refuse to support international law when there is so much to be gained and so little to lose?  With the anti-weapons law, all a nation has to give up is the perceived right to possess armaments that can annihilate the people of other nations—a perceived right that surely will lead to the eventual annihilation of the people of all nations. This answer, as reasonable as it seems to me, does not satisfy everyone.  So purely for the sake of argument—because I cannot imagine that such a development ever would occur—let us suppose Japan breaks step with us.  We would impose an embargo on manufactured goods that would apply to Japan and nations trading with Japan.  The result would be hardship for Japan and its trading partners.  To be perfectly realistic, it also would present problems for us.  However, Japan and those trading with it would suffer far more than the United States.  They would be cut off from a rich and stable market for manufactured products.  They would be denied access to the sophisticated and well-to-do customers who purchase automobiles, televisions, VCRs, microchips, and other advanced products.  Japan's loss of U.S. customers could not be offset by aggressive marketing in less affluent nations.

The United States, in turn, would lose markets that are important to it.  Some segments of the American economy—computer software, for example—would be especially vulnerable.  But other segments might see the withholding process in terms of opportunity.  We Americans well might begin to ask why we cannot start building our own VCRs, why we cannot reinvigorate our consumer electronics industry so that it can start meeting our own requirements and perhaps again compete in world markets. The United States has a vast reservoir of competence and ingenuity.  I believe we can meet the economic challenge that might result from our leadership in the cause of world preservation. Certainly, there will be difficult times for companies, individuals and the total American economy, but our country has survived other difficult times and we have grown stronger in the process.

The important thing for Americans and the citizens of other nations to remember is that the prize is worth the price.  Even if domestic manufacturers failed to provide an adequate, reliable source of the manufactured goods Americans now import,  even if we had to deny ourselves the products made by Japan, would not the hardship be worthwhile?  Isn't a human life worth more than a VCR?  Isn't the preservation of human society more important than a portable sound box? So instead of concentrating on possible difficulties, on the problems we might have to face, we ought to take a fresh look at what we seek to achieve. We ought to consider some of the developments of recent history. We ought to ask ourselves how they might have been influenced by the existence of an international anti-weapons law, one supported by two-thirds of all the world's nations, including those doing two-thirds of all the world's commerce.

Given the existence of such a law, would Iraq and Iran have been locked in an eight-year death struggle that has inflicted unspeakable suffering on both soldiers and civilians?

Would Iraq have resorted to the chemical weapons that civilized society renounced seven decades ago?

Would Iran have turned the Persian Gulf into a war zone, endangering the safety of the ships of even neutral nations?

Would Cuban troops have been committed to the defense of the communist regime in Angola?

Would the Soviet Union have invaded Afghanistan?

Of course, there is no way to be sure of our answers to these questions, no way to precisely determine what might have been. But it makes no sense to totally deny the possibile effect of enforceable international law. It is not reasonable to suppose that the world's most powerful nations, including the two superpowers, could not influence behavior within the international community. Acting in concert in support of a non-violent solution to the problem of aggression, the world's most powerful nations most assuredly could achieve a great deal. At the very least, they could make important progress in the quest for lasting international security. If an international anti-weapons law had been in effect over the past decade, it is highly probable that much bloodshed would have been avoided. If the exporting of terrorism had been banned by world law, many families would have been spared agony. Under the protection of the rule of law, our planet would be a far safer place. Trade—the great civilizing power—would become the great peacekeeping power, the power that would ensure lasting and genuine protection from the barbarism of aggression and terrorism. Then humanity, at long last, would be free to achieve its vast potential.

# PREMISE III

## CREATING
## ENFORCEABLE
## WORLD LAW

A world law would go into effect when it is adopted (1) by two-thirds of all the world's nations and (2) by nations doing two-thirds or more of all international commerce in manufactured goods. This formula, which must include the Soviet Union, assures persuasive political and economic support for the law and its objectives.

# 9

# A Moment Of Risk,
# An Age Of Agreement

*We must create world-wide law and law enforcement as
we outlaw world-wide war and weapons.*
—John F. Kennedy

When the question of an enforceable international anti-weapons law achieves a place on America's agenda, the debate will be intense. The Congress and the White House, politicians and pundits all will have their say. The policies of the past will be attacked and defended. Plans for the future will be praised and denounced. But as the arguments continue, U.S. policy will begin to change. Our national leaders will have heard the voices of people who are concerned about the continuing existence of earth-threatening armaments and the serious economic problems these weapons have helped to create. The national leadership will have begun to move further in understanding the necessity of turning away from international stalemate, of embarking on a bold new course of action. This change of direction will not be radical at first. Discussion and compromise—cornerstones of democratic society—do not lend themselves to abrupt shifts in guiding principles. But subtly, carefully, the United States will lead the world toward a new level of peaceful, productive cooperation.

To begin with, the United States and like-minded nations will move with caution toward the development of a system of non-provocative defense. This will call for a substantial reductions in the development, production and deployment of weapons of mass

destruction.  The program also will call for the complete elimination of the sale of such weapons to other nations.  Concurrently,
the United States and the nations joining it will make clear their
continued opposition to aggression.  They will maintain, and perhaps even strengthen, their defensive capabilities and they will
assert their right to supply defensive arms to any nation facing the
threat of armed aggression.  In other words, the nations seeking
to establish lasting international security will not be repeating the
mistakes of Munich.  They will not be endorsing a peace-at-any-
price philosophy.

By reducing their investment in new and redundant systems for
the obliteration of the world's civilian population, the United
States and like-minded nations will not be abandoning the cause
of freedom.  They will be moving toward a day when substantial
reductions in overall arms expenditures will become a reality,
reductions which will free economic and intellectual resources that
can be rechanneled into activities that will serve the best interests
of people in every corner of the world.  The announcement of
policies that represent a shift from the long-standing reliance on
weapons of mass destruction will receive mixed reviews in many
nations.  In the European Community, for example, we can expect
some expressions of concern that the United States is abdicating
its responsibility under the NATO agreement, that it is opening a
door that will invite an unstoppable takeover by conventional
Soviet forces.  On the other side of the world, at the other end of
the Soviet colossus, China, Korea, Taiwan, Thailand, and Japan
may share the fears of some French, West Germans, British and
others.  Leaders of the Pacific nations, long protected by U.S.
armed might, may feel they run an increased risk of renewed
Soviet expansionism.

Indeed, the proposed new policy does involve some risk.  But
nothing of great value can be obtained without risk.  It is an
inevitable consequence of leadership.  By facing up to risk, we can
gain the Fourth Freedom for coming generations.  By investing a
limited period of uncertainty and accepting the possibility of a
modest amount of economic discomfort, we can purchase limitless
years of security.  By any standard of judgment, the transaction
will be one of history's greatest bargains.  Since the new policy is
nonconfrontational, nondiscriminatory and offers all nations substantial benefits, the level of risk will be low.  In the end, the
concept of defense based on economics will prevail—not necessarily because of universal agreement but because the U.S. economy,
freed of excessive armaments expenditures, will be reasserting

itself as a creative driving force for the economy of the world. Self-interest will motivate many nations to join the United States in leading the world away from disaster.

In the Soviet Union, the new policy will be carefully analyzed. Through secret discussions within the Politburo, military, economic and political options will be evaluated. There is even a chance that these discussions will lead to endorsement of the initiative, to a superpower partnership approach to the development of a new and effective system of providing genuine international security. But judging by past experience, the most likely initial response by the Soviet Union will be some kind of propaganda attack on the United States and nations that join with it. Despite the changes in Soviet policy advocated and implemented by President Gorbachev, it is not reasonable to expect immediate Soviet endorsement of an initiative so clearly identified with the United States. Decades of insularity and the tradition of suspecting U.S. intentions and U.S. motives will continue to dominate the thinking of at least some segments of Kremlin leadership. Besides, a continual state of war—even a cold war—is not entirely a bad thing for leaders who want to protect their prerogatives and position. Lenin, Stalin and their successors all took full advantage of real or imaginary threats to the Soviet state. While the world hopes this chapter of Soviet history is closed, Soviet leaders of the present generation are unlikely to totally abandon the traditions of the past by quick acceptance of a proposal originated by the United States and its democratic friends. The leaders of the Soviet Union also will be counting on defections by U.S. allies and objections by non-aligned nations. The initiation of CDP will present the Soviets with an opportunity to use their propaganda to separate the United States from the rest of the world.

Within America, opponents of CDP will conjure up a frightening picture. They will warn that the United States will be cut off from the mainstream of the world economy and that U.S. trading partners in Western Europe and the Pacific basin will gravitate toward the economy of the Soviet bloc. My own experience in world trade tells me such a prediction is out of step with reality. In the first place, we have to remember that phase I economic sanctions, the kind we are talking about here, apply only to manufactured goods. Earth products are exempt and trade in earth products would not be affected. A second point that needs to emphasized again is that the United States is a vast and attractive market for the manufactured goods of other countries, a market for which there simply is no adequate substitute. The United States also has enormous natural and productive resources. It can get along without the manufactured goods it imports from one, or even more than one, of its trading partners. Under the worst situation imaginable—abandonment by all of the countries with which we trade—we would be a long way from total collapse.

On the other hand, nations refusing to join with the United States in a crusade to eliminate weapons of mass destruction would invite serious economic difficulties for themselves. They would be cut off from trade with the United States and with other nations adopting the U.S. position. Turning to the Soviet Union and its satellites would be of little help. The East European bloc provides a relatively poor market for the kinds of goods the United States imports and produces little of the kinds of goods the United States exports. Given the realities of national self-interest, the traditional trading partners of the United States should have no difficulty in deciding to endorse the initiative. Indeed, I would venture the opinion that some of our trading partners will support CDP more quickly and enthusiastically than many Americans.

Nations like India, which enjoy apparently cordial relations with the Soviet Union and which trade with both East and West, certainly will think long and hard about the wisdom of isolating themselves from the world's most productive economies. Even Soviet client states will face a dilemma. For example, would the Cubans really want to be more dependent on the Soviet Union than they are at present? While other nations consider their options, U.S. leaders will be under intense internal pressure to relax their position. The resolve of our nation will be subjected to a severe test. We will have two choices: We the people can demand the softening or abandonment of the U.S. stand, in effect returning civilization to a state of continuing danger, or we the people can insist that the United States move forward with a program that will free the world from fear and allow civilization to flourish. If the issues are presented through rational debate and if the debate is reported by the mass media with at least a degree of objectivity, the American people will reject the past and come down hard on the side of the future. They will give their government the support it needs to proceed with additional steps in the implementation of the Civilized Defense Plan.

The first of these steps is a proclamation of this kind:

> To secure for ourselves, our posterity and all humanity freedom
> from nuclear and other civilian life threatening weapons, we will

suspend trade in manufactured goods as of (date) with nations that use, threaten to use, supply such weapons to others or commit acts of armed aggression. We also will suspend trade in manufactured goods with any nation that continues to trade with nations guilty of armed aggression or with nations that refuse to renounce weapons of mass destruction. We invite other nations to join us in setting in motion a process that will lead to the elimination of weapons of mass destruction, reduced expenditures for defense, expanded global economic opportunities and lasting, enforceable international security.

The declaration will come as no surprise to the nations of the world. They will have been informed of U. S. intentions and they will have been asked to cooperate by adopting the declaration as their own national policy. By constantly and accurately communicating its intentions, the United States will be paving the way with other governments for support of an international weapons law. The law will go into effect when it is ratified by two-thirds of the nations of the world, including those doing two-thirds or more of all international trade in manufactured goods. As I've noted, the ratification formula also requires the inclusion of the Soviet Union.

> **Putting the responsibility for ratification of an international weapons law on the shoulders of the industrialized nations is a recognition of reality. It is the industrialized nations that have brought the world to the brink of the nuclear abyss.**

The formula is squarely based on the necessity of cooperative action. It requires the involvement of an unmistakable majority of the industrialized nations, the nations that produce much of the goods and services needed by the world, the nations that can effectively enforce an embargo. Putting responsibility for ratification of an international weapons law on the shoulders of the industrialized nations is a recognition of reality. It is the industrialized nations that have brought the world to the brink of the nuclear abyss. By pulling back, by eliminating the weapons of mass destruction, they can make a profound contribution to international stability and prosperity. They can create conditions that will benefit the least developed as well as the most developed nations.

The developing nations may have the most to gain from the

elimination of weapons of mass destruction. This was the message of the late Indira Gandhi, India's prime minister, in a statement to a 1981 conference of the United Nations Food and Agriculture Organization. She said that for the price of one intercontinental ballistic missile it would be possible to "plant 200 million trees, irrigate one million hectares, feed 50 million malnourished children in developing countries, buy a million tons of fertilizers, erect a million small bio-gas plants, build 65,000 health care centers or 340,000 primary schools."

Some developing countries are understandably indifferent to disarmament issues. They are more concerned about internal problems—hunger and unemployment, for example—that appear more immediate and dangerous than a nuclear attack. Because the economies of these nations are in a primitive state of development, it is unlikely that they initially would actively promote CDP. Nations struggling for survival and heavily burdened by debt have little to fear from an embargo of manufactured goods. But while CDP's stick might prove ineffective, its carrot can do the job. Developing countries could be enthusiastic supporters of CDP if they understood its many benefits—the economic, technical and other incentives that would be tied to the creation of a system of international security based on economic rather than military power. The other side of the coin, the economic pressure the United States will threaten to impose in support of the anti-weapons law, will be as humane as possible. Earth goods are not included in the phase I embargo because they are essential to human survival. Prohibiting their trade at this stage of the Plan's implementation would punish the ordinary citizens of a country, increasing tensions and threats of reprisals while having only a marginal effect on the near term capacity of a country to wage aggressive war. For the same practical and humanitarian reasons, health care supplies, including medication and equipment, also would be exempt during phase I.

The focus of the declaration is on the industrial products that are essential to the development and maintenance of an arsenal of nuclear, chemical or biological weapons. I again must emphasize that in today's global economy, no nation can be completely self-sufficient in the products of industry. Even the United States with its strong and diversified economy would experience hardship if manufactured products were withheld from it. For most other nations, the withholding of these products would strike a devastating blow at the economy, make aggressive behavior undesirable for the most pragmatic of reasons and encourage ratification of the law. After the law goes into effect, civilized nations will be able to utilize the phase II embargo to enforce a world-wide prohibition of weapons of mass destruction. The phase II embargo calls for total economic isolation of nations violating the anti-weapons law. The embargo would be imposed against both law-breaking nations and their accomplices, that is, nations that continue to trade with the lawbreakers, and it would include both manufactured and earth goods.

While the nations of the world are in the process of adopting the anti-weapons declaration, of creating a new and binding international law, they also will be negotiating the details of implementation. The first step will be reexamination of the definition of weapons of mass destruction that was included in a 1948 resolution adopted by the United Nations. The definition, cited in the introduction of this book, includes ". . . atomic explosive devices, radioactive material weapons, lethal chemical and biological weapons, and any weapons developed in the future which have characteristics comparable in destructive effect to those weapons . . . ." The definition is broad. But the national leaders involved in CDP's implementation must make sure that it covers every possible instrument of mass terror, that it takes into consideration the technological advances made since 1948 and the breakthroughs that are likely to occur during the years ahead. In reviewing and perhaps revising the definition, there is bound to be much discussion and perhaps some compromise. But with the promise of security so near at hand, basic principles will prevail. National leaders will realize that their defensive capabilities will not be impaired. They will not be at the mercy of neighbors motivated by greed or ancient grudge. In fact, defensive capabilities will be more equal than they were during the decades from 1945 to the start of the CDP era. The nuclear club will have closed its doors forever. Every nation will be in a position to maintain an appropriate defense, one that provides adequate protection without threatening other members of the world community.

There is nothing wrong with maintaining an appropriate level of defense. Every living creature has a defensive system, a fact underscored by the very existence of the creature. On the human level, the existence of individuals and nations depends on their ability to defend themselves. As a Hindu proverb puts it, "A man without a stick will be bitten even by a sheep." With CDP in place, nations no longer will have to burden their economies with the costs of defensive systems based on the questionable doctrine of retaliation. They will be able to concentrate their resources on the development of mechanisms designed solely to protect them against aggression.

Of course, CDP will not eliminate ethnic and religious hatreds or rivalries between nations. But CDP will shift international competition from the battlefield to the production plant and the trading floor. It will force contending ideas to vie for supremacy in an environment that is free of violence. As support for the weapons law develops around the globe—as it surely will—the Soviets will be faced with a difficult decision. They will realize that the law represents a fundamental change in relations among nations. Unlike a period of detente, which can disappear as rapidly as it developed, CDP involves a permanent reshaping of international affairs. Aggression for territorial, ideological or any other reason no longer will be tolerated by civilized society. So after their propaganda barrage fails to win significant support among nations, the Soviets will have to choose from three alternatives, two filled with risk, the third with promise:

They can roll the dice by moving quickly to satisfy longstanding territorial and ideological ambitions, an action that could trigger a cataclysmic response by the West.

They can continue to pump their wealth into the bottomless pit of arms expenditures.

They can affirm their place in a civilized society by accepting the international weapons law.

A decision of acceptance—the move that appears more likely than ever considering the most recent U.S.-USSR agreements— would end the threat of war as we have known it in this century. If Soviet leaders accept the law, all they stand to lose is a millstone. They will be hailed for their leadership in the cause of peace and they will be free to continue to proclaim the superiority of Leninism to all who will listen. Best of all from their point of view, they will have access to the technology their nation desperately needs.

The concept of open trade in all manufactured goods, including products involving the most advanced technology, should be particularly appealing to Soviet leaders. With weapons of mass destruction eliminated, the need for secrecy will be greatly diminished. The United States and other producers of manufactured goods, especially high tech equipment, will be free to compete for Soviet business, a competition that could convert the Soviet Union into a major market for a wide range of manufactured products.

Given the alternatives, the Soviets will be realists. It seemed impossible not long ago and, to my mind, it remains somewhat unlikely at present, but they could be among the first to ratify the law barring weapons of mass destruction from the face of the earth.

# 10

# A Formula
# For Unlocking
# The Future's Potential

*The world in arms is not spending money alone. It is spending the sweat of its laborers, the genius of its scientists, the hopes of its children.*
— President Dwight D. Eisenhower

When I discuss the formula for adoption of CDP with others, I often am surprised by their initial reaction. Americans in government, academia, law, business and many other walks of life, even those who are part of the peace movement, express considerable skepticism. Some see all kinds of obstacles. Others

> The naysayers are afraid that people are not the same the world around. They are afraid the people of other nations do not want the same things we want for ourselves and our families.

say flatly that the formula won't work. When I ask them why a formula based on the two-thirds principle so universally accepted by democratic society won't work, I receive a variety of answers. But in the final analysis, all are based on fear. The naysayers are afraid that people are not the same the world around. They are afraid the people of other nations do not want the same things we want for ourselves and our families.

My experience has taught me a different lesson. I believe peo-

77

ple of varied national identities and cultural backgrounds are fundamentally the same. They need food, clothing and shelter. They want the opportunity to earn a living, utilize their talents and practice their religion. They also want to be free of the threat of devastating war. This desire, shared by people around the globe, explains my faith in the eventual affirmation of the CDP concept. If the people of a nation want something strongly enough, their leaders sooner or later will have to grant it. When the people and their leaders recognize that CDP has the potential to free the world from fear, the adoption formula will offer a common sense way to bring the potential to reality. By requiring international law to be adopted by two-thirds of all the world's nations, CDP assures that the law will have a strong and lasting base of support. A simple majority, one that can disappear with the defection of a few nations, is not enough for a law that is so important to all the world's people.

There is nothing magic about the two-thirds formula. But it does have a long and honorable tradition in democratic society. It has been used around the world as a standard of decisive popular or parliamentary support in matters of great consequence. The founding fathers of the American Republic included it in a Constitution which has endured for more than 200 years. The two-thirds formula of support has proven itself again and again. It is useful, workable, understandable and effective. Common sense dictated its place in CDP. In CDP, the two-thirds formula has both political and economic dimensions. To achieve strong political support of the anti-weapons law, the formula requires adoption of the law by two-thirds of all nations. To assure the powerful economic support that is necessary for the law's enforcement, the formula calls for endorsement of the law by nations doing two-thirds of the world's trade in manufactured goods. Common sense mandates the inclusion of the superpowers in any effort to achieve international security and the two-thirds formula does just that. Common sense also requires support by the industrialized nations which are largely responsible for today's threat to the earth's survival and by those nations still striving to develop their economies. To provide the protection that the world longs for, the anti-weapons law must have the backing of nations rich and poor. It must enjoy the support of an unquestioned majority of nations and a preponderance of the world's manufacturing capacity.

But another question persists: Why can we expect any nation to adopt a law that breaks so sharply with established, if obviously inadequate, concepts for achieving international security? The most apparent answers, given earlier in this book, involve a host of benefits: Freedom from the threat of annihilation, the end of international aggression, reduction of the burden of defensive armaments, elimination of destabilizing state-sponsored terrorism and, perhaps most important of all, economic stimulation through the lowering of trade barriers around the globe. But there is another, perhaps more fundamental answer that has to do with humanity's future. One way to understand this answer is to look at the world as it is today. It is a world in which security issues seem to be addressed haphazardly. Minor questions take on major importance. In a world without an enforceable security system, nations deplete their energies dealing on an ad hoc basis with rapidly changing situations and conditions.

From the American perspective, an excellent example is provided by our relationships with such nations as Iran, Iraq, Nicaragua, Panama, and Libya, nations that, by most standards of measurement, cannot be ranked among the great powers of our age. But despite their limited size and relatively weak economies, they consistently have captured the attention of the White House, the Congress and the American people. Iran, Nicaragua and Libya, in particular, have been the subjects of endless policy debates. The question of how we should handle these nations has exacerbated tensions between the Executive and Legislative branches, supplied the raw material for countless political speeches, complicated our relationships with the industrialized democracies, the Soviet bloc and developing nations and, when all is said and done, left American public officials and private citizens feeling frustrated and impotent. The great American superpower has been unable to develop a consistent response to the problems created by nations that, with all due respect, have to be assigned to the second or third levels of international importance. And the process of seeking a response has drained our intellectual, emotional and material resources. It has prevented us from focusing our national energies on problems that clearly are within our power to resolve.

But suppose for a moment that CDP was in effect. The questions of Iran, Nicaragua, Libya and other nations no longer would have to be answered unilaterally by the United States. The United States no longer would have to play the role of global policeman. Questions dealing with the behavior of a nation would be dealt with by the international community within the framework of international law. In the case of Iran, its attacks on international commerce in the Persian Gulf and its support of terrorist activities in the Middle East and elsewhere well might have brought about imposition by civilized society of a total trade boycott, the ultimate weapon against an outlaw nation. The same penalty also might have been imposed on Iraq, Iran's enemy in the Gulf war, for the invasion that appears to have triggered the war and for the well-documented use of outlawed chemical weapons against defenseless civilians. With the boycott in effect, Iran and Iraq would be denied access to the materials needed to wage war or even to maintain a civilian economy. Both would have to rejoin the civilized world by renouncing their criminal activities. The alternative inevitably would be economic disaster and, ultimately, political upheaval.

Nicaragua's economy is even more fragile than those of Iran and Iraq. If the actions of Nicaragua's leaders—for example, the sponsorship of terrorism in El Salvador—led to the imposition of the total trade boycott called for in phase II of CDP, collapse would be swift and complete. Under CDP, the Soviet Union, Cuba and, for that matter, the United States and its allies would be barred from exporting aggression. Treaty relationships could be maintained by the East and the West. But no longer would it be permissable for the nations of either bloc to intervene militarily in the affairs of others. Cuban and South African troops would have to leave Angola. Agents of the CIA and the KGB no longer would be permitted to organize and finance armed aggression. No nation would be allowed to mount an attack on another without suffering the penalty of economic isolation.

In the world of CDP, armed aggression would be self-destructive and every nation would reap the rewards of mutual protection. For the first time in human history, governments would be free to concentrate an overwhelming share of their resources on the enhancement of human life. CDP has the power to eliminate many of the fears that for centuries have eroded relationships among nations. But I ask you to bear in mind that CDP does not have the power to create a perfect world. CDP cannot eradicate every trace of gangsterism or adventurism. It cannot—and should not—snuff out the aspirations for autonomy of restive ethnic groups. It cannot guarantee the elimination of every act of international terrorism.

What CDP can do is establish the primacy of the rule of law in international relations. The benefits are self-evident. Civilized people accept the necessity of law in their day to day lives. Families live side by side in relative harmony. They do not fortify their homes out of fear that they will be invaded by their neighbors. They know acts of violence are forbidden by law and they have faith that the law will be enforced. It is far cheaper—and far more civilized—to maintain a criminal justice system that serves an entire city than for each family to seek out and punish those it feels have transgressed against it. By working together, people achieve their common goals, strengthen the fabric of society and, at the same time, conserve resources that eventually may be needed for solving other problems.

The law enforcement experience of the world's cities certainly can be applied to the world itself. Instead of worrying about their own defense, the nations of the world ought to think in terms of providing for the common defense. In this way, they can defend themselves against would-be outlaws and prepare themselves for a range of challenges that already are on the horizon. In *State of the World—1987* (W. W. Norton & Company, New York, New York, 1987), Lester R. Brown and his fellow researchers at the Worldwatch Institute list a few of the challenges involving what the researchers describe as "our relationship with the earth and its natural systems . . . ." For example, population growth is outstripping natural resources, especially in many of the least developed nations, contributing to runaway urbanization and dramatic increases in poverty, hunger and disease. In most parts of the world and especially in the United States, waste disposal has become a major problem. Landfills are reaching capacity, incinerators can add to airborne pollution and only token attention is being given to the recycling of valuable materials.

The list of challenges goes on: From the depletion of the earth's ozone layer to the changes in climate that inevitably will alter agricultural practices and from the post-Chernobyl reappraisal that threatens the future of nuclear power to the not fully understood die-off of trees that could change the face—and the economy—of vast sections of Europe. There is also the challenge of sickness—cancer, heart disease, mental disorders and, of course, acquired immune deficiency disease, the much-discussed AIDS problem that is having a profound effect on many aspects of life in America and other nations. While these challenges go unmet—or undermet—the world continues to pour its resources into its armies and navies. For example, *The Relationship Between Disarmament and Development* (United Nations, New York, New York, 1982) reports that ". . . the world's regular armed forces total some 25 million persons. This figure has risen steadily over the past 20 years, with net stability among the developed countries accompanied by increases in many developing countries. The global figure in 1980 was more than 10 percent larger than in 1970 and nearly 30 percent larger than in 1960."

In addition to those in the regular armed forces, the United Nations report estimates that worldwide there are 10 million persons in paramilitary units, another four million who are civilian employees in departments of defense, 500,000 engaged in research and development, perhaps four to six million workers directly involved in the production of weapons and other specialized military equipment and another three to six million industrial workers whose jobs are indirectly supported by military expenditures. So in terms of a fairly rough and probably conservative estimate, more than 40 million individuals—many of them highly trained and potentially productive—currently are part of the worldwide military machinery.

How much does all this cost? In a report for 1987-88, the International Institute for Strategic Studies estimated that worldwide military spending in 1986 amounted to approximately $900 billion with the United States and its NATO partners and the Soviet Union and its fellow signers of the Warsaw Pact accounting for 75 to 80 percent of the total. An increasing share of the dollars spent for military purposes pays for sophisticated weapons which require exotic materials often incorporating minerals that are in relatively short supply. For example, titanium now is being used extensively in combat aircraft and submarines. In the case of aluminum, copper, nickel and platinum, according to the United Nations report, "estimated global consumption for military purposes is greater than the demand for these minerals for all purposes in Africa, Asia (including China) and Latin America combined."

Now there is no guarantee that disarmament would produce an automatic redirection of resources. There is no certainty that an outbreak of peace would assure the rechanneling of manpower, dollars and natural resources to meet the enormous challenges of today and tomorrow. However, one thing is certain: Unless we eliminate fear-breeding, aggression-generating weapons of mass destruction, unless we tackle the dangerous problem of terrorism, there is no hope for the future. We will either blow ourselves out of existence or, if we somehow manage to avoid that risk, we will see civilized society paralyzed by terrorist attacks.

At this moment, of course, we are trapped in a lawless world, one that is threatened by disasters of our own making. But there is a way out of the trap, a way to bring to the world what can only be described as a new birth of freedom. By following the course mapped by CDP, the world can stop squandering its treasure on weapons and increase its investment in the future. With CDP in effect, the focus of competition can shift from the military to the commercial. Resources now allocated to armaments will be available to solve problems and take advantage of opportunities. Weapons of mass destruction can be exchanged for weapons of trade.

In the United States, a share of the dollars and brainpower now employed in weapons projects could be used to rebuild highways, sewer and water systems and other infrastructure components. A 1988 report prepared for the United States Conference of Mayors by Employment Research Associates argues that ". . . transferring $30 billion per year from the military budget into urban programs can make a sustained contribution to a higher quality of life in American cities. It can mean that the nation's children are better educated, the public health system improved, the elderly are given better care, more housing is provided for the poor, and city life is made less stressful and less polluted. A welcome collateral benefit of these new budget priorities would be more jobs, more investment, more tax revenues, and a higher GNP for the nation." Dollars now used for the development and purchase of aggressive weapons also could help revitalize the industrial base and to stimulate the technological advances that would increase productivity and lead to the creation of new, life-enhancing products. Some of the new products could be based on recent breakthroughs in such fields as super-conductivity, artificial intelligence or biological engineering. Still other new products could come from industries that do not even exist today—industries that are sure to be created by the ferment of scientific and engineering activity that will occur when attention shifts from death-centered to life-centered research and development.

Of course, some jobs that depend on arms production will disappear. But in their place will be new jobs, permanent jobs, jobs less dependent on political pressures or potentially wasteful cost-plus contracts. In the world after CDP is implemented, growth in trade can help all nations come to realize how much they depend on others. All nations can begin to understand that isolation can produce devastating results, that there truly is an international community and it must consciously seek the strengthening of commerical, cultural and all other connections. The benefits of an interconnected world have been demonstrated by the experience of the European Community. As fear of aggression subsided in Europe, cooperation increased, creating an economic entity that has far more clout than any of its member nations could hope to achieve independently. In 1992, the 12 nations of the European Community will become a single market and, with its 320 million people and vast economic resources, it will be the richest market in the world.

In North America, the United States and Canada long have shown what can be accomplished when the threat of military aggression, implied or explicit, does not exist. The border between the two nations is not fortified. Neither nation has missile batteries programmed to destroy the capital of the other. Peace has created a trading relationship that is so mutually beneficial that neither Washington nor Ottawa can afford to give any thought to settling differences through military action. Nearly 200 years of peaceful U.S.-Canadian trade, the more recent history of European economic cooperation and the spectacular achievements of Japan, the quintessential trading nation, provide an indication of some of the benefits that are possible when international relations are reordered by the substitution of economic for military power.

> **By moving the focus of world attention from the balance of terror to the balance of trade ... we can create an environment that fosters the productive relationships that serve the interests of all nations and all people.**

With its security safeguarded, no nation should have reason to fear the reality of global interdependence. People should begin to understand that the economy of each nation depends on the economies of all nations. Those in the largely developed nations of the Northern Hemisphere cannot achieve enduring prosperity as long as the underdeveloped nations of the Southern Hemisphere are impoverished. There is good reason to believe that increases in the volume and quality of international trade will help strengthen the economies of developing as well as developed nations. The developed nations need markets for advanced products. The developing nations also need markets for the goods they are capable of producing. In addition, they need the technology and expertise that are only available through trade. By moving the focus of world attention from the balance of terror to the balance of trade, we can make war unthinkable. We can create an environment that fosters the productive relationships that serve the interests of all nations and all people.

Political economist Lloyd J. Dumas, who I referred to earlier in this book, described such an environment in a 1987 lecture at the University of Notre Dame: "What we need is an economic system where there is too much to lose by going to war, a system that has benefits flowing in all directions." The goal of CDP is to bring such a system within reach, to create a world in which nations are free to concentrate their attention on issues of human survival and fulfillment. Distractions and irritations—how to deal with an Iran or an Iraq, a Nicaragua or a Libya, for example—would be minimized. Dead-end armaments expenditures would be reduced. The economy of the world would be given an opportunity to meet the needs and expectations of an expanding global population. This is why nations will move to adopt the international law called for by CDP. This is why adoption by two-thirds of the world's nations is a reachable goal. If the world's citizens and their leaders understand what can be gained through CDP, they will have little hesitation about their endorsement. They will accept the Plan with enthusiasm and they will extend their full cooperation to the international monitoring and enforcement mechanism that we next will examine.

# PREMISE IV

## Establishing
## An International Agency
## To Monitor Compliance
## With The Law And Mobilize
## The Economic Power
## Required For Effective
## Enforcement

For enforcement of the law banning weapons of mass destruction, supporting nations would establish a limited purpose international agency with the authority and ability to define, detect and dispose of such weapons presently in existence, prevent further manufacture of such weapons, identify acts of armed aggression, monitor trade with outlaw nations, marshal world opinion against aggressors and issue the call for imposition of economic sanctions.

# 11

# Policing The World With Non-Provocative Power

*We must break out of the straitjacket of the past. We must have sufficient imagination and courage to translate the universal wish for peace—which is rapidly becoming a universal necessity—into actuality.*

—Douglas A. MacArthur

Law is a mark of civilization and civilized society always has understood that law enforcement cannot be left to the individual. It must be a responsibility of some governmental agency and, to be truly effective, it must reflect the values of those it has the power to affect. All this ought to be self-evident. Yet, on the international level, society has taken only minor steps to develop a global law enforcement system. The nations of the world—like the barons of Europe's dark ages or the Chinese war lords of the early 1900s—have been reluctant to share power with others, even in matters involving vital common interests.

> **Civilized society has not created a system for protecting itself . . .**

So while there are international agreements that are acknowledged as having the force of international law, compliance has been voluntary and enforcement has been unreliable or nonexistent. A recent example involves the use of poison gas by Iraq. This relatively small nation clearly violated an agreement that has

89

been in effect for 70 years—and got away with it. Of course, some members of the international community protested. But Iraq knew it could ignore the protests. What did it have to fear in a world without an effective mechanism for the enforcement of international law? The world's Iraqs know they can behave in the most outrageously uncivilized manner without any worry about punishment. Civilized society has not created a system for protecting itself from nations that break the law. Nations are reluctant to share police powers with other nations or with some central world authority. They are suspicious of the motives and intentions of others. They fear what might happen if some of their power fell into the hands of a hostile nation or group of nations.

Part of the problem involves the concept of power. In conventional thinking, power means physical force. It signifies armies and navies, ships, planes and tanks, weapons that can be used with devastating effect against a nation and its people. No nation is likely to voluntarily share a significant amount of this power with other nations, at least not with nations that are real or potential enemies. But there is another kind of power, the kind that has been emphasized repeatedly on these pages. By substituting economic power for military power, the world can reduce the fears of law abiding nations. The world can create an international law enforcement mechanism that will not be a threat to the integrity of national borders. This non-provocative agency will have the power to command respect and obtain compliance with the anti-weapons law but not the power to achieve international military dominance.

The term "non-provocative" is important. The new international monitoring and enforcement agency in no way will be capable of aggressive action. It will not have the authority, equipment or personnel to threaten the territory or people of any nation. Instead, the agency will play a major role in eliminating the aggressive capabilities of all nations. The goal will be a world in which nations will be able to protect themselves but will not be tempted to attack others. As President Franklin D. Roosevelt noted in a 1933 statement to the Geneva Disarmament Conference: "If all nations will agree wholly to eliminate from possession and use weapons which make possible a successful attack, defenses will automatically become impregnable, and the frontiers and independence of every nation will become secure."

Because it will be non-threatening, the new international monitoring and enforcement agency will be much more attractive to the governments of the world than an armed international police force. An armed international police agency is a threat and the world's leaders are unlikely to replace old threats with a new one. Besides, it is inconsistent and counterproductive to use war-like means to eliminate war, a lesson we should have learned from the "war to end wars" of 1914-18.

Supported by a two-thirds majority of all the world's nations, led by those with the most powerful economic systems, the new international weapons agency will be fulfilling a role once intended for the United Nations. The U.N. Charter, like the Covenant of the League of Nations which preceded it, authorized the enforcement of international law and over the years some attempts have been made to implement the authorization. Recently, for example, the U.N. played a key role in bringing about a cease fire in the Iran-Iraq conflict and U.N. observers are being utilized to oversee the withdrawal of foreign troops from Angola. But for the most part, U.N. attempts to enforce world law have been frustrated by Security Council vetoes.

The veto system gives the Security Council's five permanent members, the United States, the Soviet Union, Britain, France and China, the power to reject any enforcement action. This power, often used in support of narrow national self-interest, has paralyzed the U.N. peacekeeping function. Through the years, there have been many attempts to do something about the veto, to reform the United Nations so that it can play a far greater role in the development of genuine international security. All of these attempts have failed and, in my opinion, future reform attempts also are doomed to failure. Nations, like individuals, are not likely to loosen their grip on authority, especially when they perceive that authority as providing them with an advantage.

Given the veto obstacle and the existence of a bureaucracy that over the years has established a fixed pattern for responding—or not responding—to flagrant violations of world order, the United Nations cannot be the sponsoring organization for an agency that will be entrusted with the enforcement of a law that is so important to the world and its future. As a practical matter, the new agency must be independent. It must have its own structure and agenda. This does not mean that the new agency will be a rival of the United Nations. In fact, the two organizations should work in close cooperation. The new organization might even have the status of a United Nations affiliate. But it must stay free of the system that has prevented the United Nations from effectively enforcing international law.

My experience in business also has taught me that the successful executive knows how to divide responsibility. The objective is
to focus maximum attention on a specific opportunity or problem.
If the responsibility for enforcing the anti-weapons law was turned
over to the United Nations, it is doubtful that the required focus
could be attained. Under U.N. administration, it is most unlikely
that the law would be given the high priority treatment it certainly would receive from an agency created solely and specifically for
enforcement purposes.

This monitoring and enforcement agency will supplement, not
abrogate or duplicate, the work of the United Nations. The
agency's duties will be confined exclusively to supporting and
enforcing the law aimed at weapons of mass destruction and military aggression. The agency will be authorized to take possession
of all outlawed weapons and to see that they are permanently
destroyed according to a schedule established through international agreement. The agency also will have the responsibility
of refining the definition of civilian life threatening weapons. The
agency will detect violations of the law, identify acts of armed
aggression, monitor trade with those nations that decline to
support the law, mobilize world opinion against aggressor nations
and, when necessary, issue the declaration that will trigger the
imposition of the total economic sanctions called for in phase II
of CDP.

The technology to carry out this work program is available.
The intelligence agencies of the United States, the Soviet Union
and other nations have developed the skills and the equipment to
keep a close watch on potentially threatening activities in every
part of the world. Through photo reconnaissance, infrared sensors, radar, high-energy particle generators, and other techniques,
each of the superpowers knows much about the weapons of its
rival. Because of the requirements of the INF Treaty, the Americans and the Soviets are even beginning to share some of their
expertise in the detection, analysis and evaluation of nuclear tests.
The INF Treaty also has brought American and Soviet observers
into installations that once were top secret. This cooperation,
while limited in scope, is a promising development. If the United
States and the Soviet Union permit their secret installations to be
penetrated by observers who are in a sense enemies, why would
they not permit the same kind of penetration by observers who
are the neutral representatives of an independent international
agency? In 1988, the doors of secrecy were opened in the name
of national self-interest. Both the Soviets and the Americans

perceived that the INF Treaty was something of value that deserved to be implemented. In future years, the Soviets, Americans and others will perceive that the international weapons law is supremely valuable and they will do what is necessary to make certain the law is enforced.

Through the cooperation of the major economic powers, the most advanced technology will be available to the monitoring and enforcement agency. Orbiting satellites, seismic monitors and other land and space based devices will be able to pinpoint the most secret weapons projects. The findings will be communicated instantly to a command center that will be empowered to set in motion appropriate enforcement actions. The automated surveillance devices will be augmented by on-site inspectors from the monitoring and enforcement agency. Like the policeman on the beat, these inspectors will carry the full authority of the law, in this case, a law that has broad international support.

> **Does national sovereignty include the right to threaten the continuation of life on this planet?**

The inspectors, like the agency itself, will not infringe on national sovereignty. If such a question is raised, the best answer involves two other questions: Does national sovereignty include the right to incinerate innocent men, women and children? Does national sovereignty include the right to threaten the continuation of life on this planet? A family has the right to defend its home from hostile neighbors. It has no right to burn its neighbor's house to the ground and to slaughter all its inhabitants. The anti-weapons law and the agency that will enforce it offer a practical way to protect nations and their people. But practical as it is, the way will not be easy. It is too new, too unconventional to be accepted without question. National leaders will have to be assured and reassured that the new international monitoring and enforcement agency can succeed in verifying compliance with the anti-weapons law. They will have to be convinced that cheating can be prevented, that weapons of mass destruction will be banned from the face of the earth.

If there is serious doubt about the agency's capabilities, CDP cannot expect to achieve broad international support. If there is uncertainty about the ability to detect and eliminate forbidden weapons, few nations would be willing to stick their necks out by endorsing a law that would require them to lay down their own weapons of retaliation. So questions about verification must be given thoughtful answers and those answers must involve not just the nuts and bolts of monitoring and enforcement procedures but the fundamental concepts underlying the agency that will put the procedures into motion.

One of these concepts has been described with great accuracy as mutually assured survival. With the acceptance of the anti-weapons law and the establishment of the enforcement agency, the universal desire for security will be achieved through reliance on defense rather than on fear of nuclear attack. Trust will begin to replace suspicion as the attitude that dominates international relationships. Nations will start to understand that their goals can be achieved without military adventurism. There no longer will be any incentive for a nation to invest in weapons of mass destruction. With the law and the agency in place, why would any nation want to run the risk of being a violator? Why would any nation jeopardize its economic future by trying to build or conceal an outlawed weapon?

Certainly no nation governed by rational men and women would challenge a law that clearly is in the best interests of all. No nation that perceives the law's many benefits would set about to undermine it. Of course, rationality and perceptiveness are not always in abundant supply in the leadership of nations. Madmen have been in charge in the past and present and it is reasonable to expect other madmen to appear in the future. Some of these well may attempt to violate the law by building or concealing a banned weapon with the intention of using it to threaten or attack a neighbor or an enemy.

With most of the world's industrialized nations supporting the law, it is inconceivable that the violation could go undetected. These nations repeatedly have demonstrated that this is a world without secrets. They have shown their ability to locate the weapons of others and they will bring this ability to the enforcement agency. They will join in outfitting the agency with the technology and skills needed to uncover existing weapons and to keep track of the materials and equipment required for the production of new weapons. It's worth remembering that CDP will not go into effect until the anti-weapons law is endorsed by those nations—the United States and the Soviet Union in particular—best prepared to succeed in weapons concealment. Both the U.S. and Soviet arsenals include nuclear missiles that may be launched from submarines or from mobile, land-based equipment. These weapons systems, which were designed to minimize the possibility of detection and which will be eliminated by CDP, are expensive and complex. It is most unlikely that they could be replicated in an Iran or a Libya.

But for the sake of argument, let us imagine that an Iran or a Libya acquired and used an outlawed weapon. The attack would be a catastrophe of gigantic proportions. But it would be a limited catastrophe. It would not lead to the global destruction that is the most likely outcome of a confrontation of weapons-possessing nations in the world as it exists today. The forceful utilization of economic power would make certain the eventual collapse of the nation violating the anti-weapons law. In other words, there can be no guarantee that CDP will work to perfection. But it will work. Violations of the law will be the exception, not the rule. If there is a nuclear attack—and I believe such an occurrence is extremely unlikely—it will involve only a few warheads, not thousands. The world will be wounded. But it will survive.

> **Economic power usually is the final arbiter of international affairs. It determines the long-term winners and losers in every high stakes contest between nations.**

CDP will work because it offers so many benefits, provides so many advantages to the nations who accept it. These benefits and advantages assure the broad support that is necessary for the effective enforcement of any law. The law that will eliminate weapons of mass destruction will be enforced by an agency that will have the multi-national backing it needs to carry out its mission, backing assured by CDP's reliance on economic rather than military power. Economic power usually is the final arbiter of international affairs. It determines the long-term winners and losers in every high stakes contest between nations. With the establishment of the rule of law on an international level and with the creation of an international law enforcement agency, economic power at last will be on the side of all who seek to free themselves from the terror of aggression.

# 12

# Bridging Differences To Establish The Rule Of Law

*Taking a new step, uttering a new word is what people fear most.*

—Fyodor Dostoyevski

For more than four decades, international relationships have been dominated by fear of the known. We have reacted to the very real possibility of nuclear war by creating a world where security rests on a fragile balance of terror. CDP proposes that we move in another direction. By utilizing economic power to establish the rule of law, CDP offers a way to break out of the prison built by fear of aggression and terrorism, to create a world filled with new opportunities for achievement. But the road to CDP is blocked by another kind of fear, fear of the unknown. We are afraid to break with established patterns, to turn our backs on accepted, but wholly inadequate, systems of international security. We also are afraid of our international neighbors and they are afraid of us. Like the cowboys in movie versions of the Old West, we are not willing to hand our weapons to the bartender until we are assured that every other cowboy will be required to do the same.

This assurance will be provided by the international monitoring and enforcement agency. But here again, we tend to be paralyzed by fear of the unknown. We have difficulty believing that the rule of law—a concept we routinely accept in our daily lives—can be effectively and impartially enforced. Part of our difficulty stems

from doubts about the mechanics of enforcement. Even though
the present generation of spy-in-the-sky satellites and terrestrial
sensors is providing the governments of the world with detailed
information on weapons development and deployment, we are
reluctant to believe in the capabilities of the next generation of
detection technology. Even though the on-site inspection procedures related to implementation of the INF Treaty have been
acclaimed as successful, we continue to be suspicious, to doubt our
ability to police the world in which we live.

> **We will finally accept that weapons of mass destruction
> cannot be concealed in a world truly committed to their
> elimination. We will understand that the mechanics of
> enforcement are well within the capabilities of society.**

Fortunately for all of us, these doubts cannot endure. They will
wither in the face of overwhelming evidence of the effectiveness
of monitoring techniques, evidence that surely will continue to
accumulate as the technology of detection constantly improves.
We will finally accept that weapons of mass destruction cannot be
concealed in a world truly committed to their elimination. We will
understand that the mechanics of enforcement are well within the
capabilities of society. But even if we concede that violations of
the anti-weapons law can be detected, there is another level of
fear that must be dealt with, the fear that the mechanics of
enforcement may be controlled by those we view as our enemies.
To overcome this fear, the world must develop something new, an
agency that is objective and impartial and, at the same time,
empowered to take prompt and decisive action. To some, the
development of such an agency is beyond the ability of society.
They cite the experience of the League of Nations and the United
Nations and, because of this experience, they see our world as
infected with an incurable disease. They concentrate on the
problem, not the solution, the disease, not the cure.
    In this case, the cure will not require any breakthrough in
human knowledge. In fact, history provides a number of examples
of systems of governance that have been successful in bridging the
differences of interest groups in order to achieve a common goal.
Our own constitutional system is perhaps the most remarkable of
these success stories. Through the process of compromise, the
differing interests of the 13 colonies were reconciled, resulting in

a governmental structure that has endured for more than two centuries. The structure was created by leaders who recognized that without some form of unity, the colonies could not hope to maintain their independence. They could not hope to preserve the traditions of personal freedom that meant so much to so many people. So the leaders of the colonies had a strong incentive to achieve compromise. They were pushed toward agreement by the political and economic realities of the 18th Century.

In the 20th Century, the pressure for compromise also comes from political and economic reality. Add to that the very real threat of world-wide devastation and the pressure becomes irresistible. The leaders of the nations supporting the anti-weapons law will establish a system of enforcement that is in harmony with the mission of the monitoring and enforcement agency, a system that will maintain its objectivity, sustain its international support and assure its permanence. Such a system must be devised by the nations supporting the law and it must be developed through the thoughtful give-and-take of negotiations that have been carefully and thoroughly planned. The system of monitoring and enforcement must not be imposed on the world by nations that are powerful and it should not be detailed in a book such as this, a book that seeks to emphasize the basic concepts that are at the heart of CDP.

However, I do feel that it is appropriate for me to outline a few of the principles that need to be considered:

1. While all nations supporting an international law must be represented in governing the monitoring and enforcement agency, the one nation-one vote system utilized in the United Nations General Assembly ought to be avoided. Giving an equal voice in world affairs to all nations—rich and poor, large and small—neither protects the interests of the have-nots nor encourages the responsible involvement of the haves.

2. The population of a nation ought to be taken into consideration in any formula for representation in the agency's governance system.

3. The economic strength of a nation, especially its involvement in international commerce and its manufacturing capacity, also ought to be given weight in deciding the configuration of the machinery of governance.

The simplest way to think of the new agency is in terms of three levels. On the first level would be those directly responsible for day-to-day enforcement activities. Here would be the technicians and others involved in surveillance, data-collection and other tasks related to the process of assuring compliance with the law. The personnel on the first level would, for the most part, be career professionals with specialized training, expertise and responsibilities. They would be under the direction of an executive with a limited term of office, perhaps no more than four years. The executive would be selected by and answerable to the agency's second or policy-making level. On this second level would be the agency's legislative arm. It would be empowered to establish the regulations necessary to assure compliance with the letter and the spirit of the law. Acting on the information and, in some cases, the recommendations provided by the executive, the agency's legislative body would refine the definition of weapons of mass destruction, certify law violations involving manufacture, possession or sale of banned weapons, identify acts of armed aggression and authorize the imposition of sanctions against law-breakers and their accomplices.

The legislative body well might be constituted according to the basic principles outlined earlier in this chapter. Under these principles, a nation's population and its volume of international trade would be used to decide its representation. The goal is to create a body that represents economic and social as well as political interests. In this body, dominating influence should be diffused across many national boundaries and not concentrated in the hands of the superpowers or of international organizations—OPEC is an example—which were created to further specific, relatively narrow interests. It also would be advantageous if the new agency started out with a charter that minimizes the possibility of entrenched officeholders. Terms of all members of the agency's legislature should be severely restricted. This restriction will help assure the integrity of the body and give it the continuing infusion of new ideas that will enable it to remain in harmony with constantly changing world conditions.

On the other hand, the agency's third or judicial level ought to be under the control of a court that is as permanent as the law it was created to interpret and uphold. It is possible that this court may become an affiliate of the World Court. But unlike the World Court, the jurisdiction of the agency's court will be limited to issues relating to the anti-weapons law and, perhaps, international laws dealing with terrorism. Most important of all, the decisions of the agency's court will be enforced by the economic power of a majority of the world's nations. The World Court's decisions frequently are ignored—even by nations like the United States—when those decisions are perceived as hostile to national interests.  But the decisions of the agency's court will carry far more weight because they will rest on the existence of an international law that has been deemed a necessity by a majority of the world's nations.  When a nation moves to support the law, it also will be giving its support to the court that is an essential part of the enforcement process.

Why is a court a necessary part of the agency?  Simply because the law is part of a Civilized Defense Plan and the concept of civilization requires a judicial system through which grievances may be heard, appeals considered, mistakes corrected and wrongs redressed.  The court offers a defense against the possibility of a misguided executive or legislature.  This, in turn, will contribute to the long-term acceptance of the agency and the invaluable contribution it can make to international security.

Because they will be composed of human beings, neither the executive, legislative nor judicial arms of the agency will be infallible.  They will be subjected to criticism that is justified and unjustified.  But even so, the agency will constitute a significant step forward for humanity.  For the first time, the world will have at its command an effective, universal process for dealing with nations that threaten the security of others.  Suppose, for example, that the agency's police force gathered evidence that Libya was constructing a nuclear weapon.  The executive would be required to immediately present this evidence to the legislative body and, along with it, a specific recommendation for appropriate action.  The legislative body also would be required to act within a specified period of time—probably no more than 24 hours—a requirement that is entirely feasible in an age of rapid transportation and instantaneous communication.  There would be time for debate but not filibuster.  If the police force and the executive did their work with competence, the legislature would certify a violation and specify a course of action involving the imposition of total economic sanctions.  The governments of the world's nations would be notified and the agency's police would begin monitoring all trade routes to make sure there is complete compliance with a civilized program designed to bring an offender to justice.

If Libya feels it was wronged, it could appeal to the agency's court.  After reviewing the evidence, the court could uphold or overturn the action authorized by the agency's legislature.  If the decision was to overturn the legislative action, the legislature could either abandon its effort or ask the executive to produce the additional evidence required to reinstate the action.  Of course, Libya could flatly reject the authority of the agency and proceed with construction of an outlawed weapon. Libya also could threaten its neighbors and, in the worst scenario, it could employ its arsenal, killing many and destroying much. But such an act of barbarism still would be far less of a catastrophe than that facing today's world.

In these final years of this century, the superpowers bristle with thousands of weapons capable of destroying all the people on earth. At the same time, renegade states take advantage of the lack of enforceable law to unleash chemical weapons on civilians and provide sanctuaries for terrorists. There is no true defense, no freedom from fear. In the world of CDP, a Libya could defy the law—for a time. But its neighbors would not be without defenses and the nations supporting the law would have the right—indeed, the duty—to provide defensive assistance to nations that are threatened. Those supporting the law, coordinating their actions through the monitoring and enforcement agency, would be able to tighten an economic noose around a lawbreaker. The lawbreaker might be able to strike back but it would be a last and dying effort, a final convulsion of contempt for civilization. With the anti-weapons law in effect and the monitoring and enforcement agency in place, civilization will survive the lunatics and bullies who rise to power from time to time. Security will become a reality for all nations and all people will have the right to enjoy the blessings of a new age of incomparable promise.

# 13

# Using CDP
# In The Fight
# Against Terrorism

*Incurable diseases are only those the doctors don't know
how to cure.*

—Charles F. Kettering

Terrorism is not a 20th Century phenomenon.  It has been used since ancient times to create the climate of fear that attracts attention to a cause, paralyzes authority and destabilizes government.   But in earlier times, terrorism was largely local.   It developed in response to local conditions.  Its announced objective was to redress local grievances.  It rarely moved beyond the walls of a city or the boundaries of a kingdom.  Then came the revolution in transportation and communications technology, the revolution that helped create the world interdependence that is basic to the CDP concept.  It now was possible for terrorist groups to operate on a global scale.  Their hiding places were multiplied and so were their targets.  And, thanks to television, their actions were witnessed by millions.

Clearly, this was a problem that called for a concerted response by the international community.  But until quite recently, much of the world reacted to terrorism with amazing tolerance.  Many of the world's opinion leaders took the position that acts of terrorism were justified by the injustices of others, that terrorism was an understandable response to colonialism, imperialism or some other perceived transgression.  Through most of the 1970s, the industrialized democracies expressed concern about terrorism but

did little to suppress it. The Soviet Union and its allies, true to
Lenin's call for worldwide revolution, denounced anti-communist
terrorism while endorsing terrorism that aimed at promoting the
communist cause. Third world nations appeared to be either
indifferent to the terrorism threat or willing to use terrorists to
achieve the objectives of the nation or its ruling elite.

But by the early 1980s, world opinion had moved sharply
against terrorism. A series of barbaric actions, including the
hijacking and bombing of civilian aircraft, the assassination, maim-
ing or kidnapping of political leaders and the prolonged captivity
of U.S. civilians and others in Lebanon, outraged civilized society,
and convinced governments that terrorists had to be suppressed.
With overwhelming public support, authorities in many nations
have taken vigorous anti-terrorism action. Terrorists have been
successfully prosecuted in the United States, Great Britain, Spain,
France, Federal Republic of Germany and other countries. The
Soviets have gone on record as opposing all terrorism and they
have specifically condemned a number of recent terrorist actions.
The United Nations, which once seemed to be the principal apolo-
gist for terrorism, has passed resolutions condemning terrorism
and the taking of hostages.

Yet, terrorism continues to be a major threat. As the continued holding of hostages, the destruction of Pan Am Flight 103
and similar acts have convincingly demonstrated, no one is safe.
The terrorist can strike at any time and in any place. Our best
security efforts have not won the battle against terrorism, they
have not succeeded in defending civilized society from the savages
who lurk in the shadows. The savages know that technologically
advanced nations are especially vulnerable. A few well-placed
bombs could destroy the electrical power distribution network,
create a communications black-out and shut down nearly all economic activity. A few ounces of a toxic substance introduced into
the water supply could create a public health disaster of unthinkable proportions.

In recognition of the increasing threat, civilized nations have
stepped up enforcement of anti-terrorist legislation and vigorously
prosecuted offenders. We have increased our efforts to keep
terrorists and their weapons off civilian aircraft, expanded our
surveillance of suspected terrorists, enhanced our ability to
penetrate terrorist groups and increased cooperation among the
world's anti-terrorist agencies. But we have been less than
decisive in dealing with those nations that support, sponsor,
encourage or condone terrorist acts. The identity of these nations

is no mystery. At the very least, the list includes Libya, Iran, Syria, Cuba, and North Korea. All have been covertly or overtly involved. All have provided sanctuaries for terrorists, plotted terrorist acts, engaged in the training and equipping of terrorists or supplied weapons to terrorists in other countries. All are guilty of behavior that cannot be tolerated by civilized society.

But the response of the United States and other nations has been inconsistent and inadequate. We launched an air strike against Libya for its alleged involvement in the bombing of a Berlin discotheque but we took no similar action against Syria, which also may have been involved in the Berlin affair and which certainly was the home base of the virulently anti-American Abu Nidal organization. We have imposed at least limited economic sanctions on Libya, Syria, Iran, Cuba, and North Korea. But this action has not resulted in their renunciation of state-sponsored terrorism. They are confident that their economic requirements will continue to be satisfied by other trading partners.

What if the United States joined with other nations in moving to implement the Civilized Defense Plan? What if we utilized our economic power to assert leadership? What if we, in cooperation with other civilized nations, brought about enactment of an international law forbidding the exportation of terrorism and then enforced the law through economic inducements and sanctions? How long could a terrorist-sponsoring state endure total economic isolation? How much time would elapse before such a state would decide it would be wise to change its ways? How much death and suffering could be prevented by resolute, responsible action by the United States and the rest of the civilized world? Unlike the military action we took against Libya, the application of CDP would not be violent. No one could question it on legal or humanitarian grounds. It would be a thoughtful and entirely appropriate response to an intolerable breach of accepted standards of international behavior.

How would the power of CDP be mobilized against the exportation of terrorism? To begin with, the United States in cooperation with other nations would have to take the lead in gaining widespread international acceptance of an anti-terrorism law. As with the proposed anti-weapons law, acceptance would be achieved through a carrot-and-stick approach. There would be economic incentives for nations endorsing a ban on the exportation of terrorism. Nations reluctant to provide such an endorsement would be faced with a threat of sanctions on trade in manufactured goods. These sanctions also would apply to the trading partners of non-endorsing nations.  In keeping with the CDP approach, the anti-terrorism law would go into effect when it is adopted by two-thirds of the world's nations and by nations doing two-thirds of the world's commerce in manufactured goods, a formula which provides the political and economic muscle needed to guarantee effective enforcement.

But what about the law itself?  Most of the behavior that characterizes terrorism already has been condemned by national law and international agreement.  Murder is murder even when the murderer carries a Syrian passport and claims to be motivated by some ideological, ethnic, religious or tribal cause.  So in increasing numbers, terrorists are being arrested, tried and punished in many parts of the world and many nations are cooperating in a growing effort to bring terrorists to justice.  Given this encouraging situation, there is little reason to complicate matters by attempting to create a comprehensive new body of anti-terrorist legislation.  What is needed is an enforceable international law dealing with a major and fundamental part of the terrorist problem: The utilization of terrorism as an instrument of national policy.  In today's world, terrorism has become a low-cost, low-risk way to wage war.  A few terrorists, adequately trained and equipped, have the potential to achieve more than an army.  They can destabilize an enemy nation, eliminate key opponents and create a paralyzing climate of fear.  For these reasons, terrorism well may become the most prevalent form of warfare during the years ahead.

> **. . . terrorism is a form of aggression and the terrorist is a potential weapon of mass destruction.**

Terrorism—at least the state-sponsored variety—can be suppressed through an international law that would enable civilized society to strike at terrorism in its lair.  Such a law would serve to mobilize the pressure that would force Iran, North Korea and others of their ilk to abandon their roles as incubators and stimulators of worldwide terrorist activity.  A number of options are available for the enforcement of such a law.  For example, responsibility could be turned over to the United Nations or one of its agencies.  But given the continued existence of the Security Council veto, the non-representative nature of the General Assembly and other shortcomings, my view is that the objectives of the anti-terrorism law could best be achieved by placing enforcement in the hands of a new limited purpose international agency, perhaps the same agency that I have suggested for enforcement of the anti-weapons law. After all, terrorism is a form of aggression and the terrorist is a potential weapon of mass destruction. So why not commission the monitoring and enforcement agency that I described earlier in this book to provide leadership in the fight against state sponsored terrorism? Given adequate authority by supporting nations, the agency could develop an acceptable definition of terrorism—a challenging but not impossible task in a society where cynics contribute to the emasculation of public opinion by glibly claiming that "one person's terrorist is another person's freedom fighter."

Armed with the definition, the monitoring and enforcement agency could move forward with its work of detecting the exportation of terrorism by one nation into another, reviewing and evaluating trade with terrorist states and, when required by circumstances, triggering the imposition of total economic isolation—the phase II sanctions called for in CDP—against those who violate the anti-terrorist law. Enforcement of the law would deprive terrorists of the sanctuaries in which they have been free to train and to which they have been able to return after a hijacking or a machine gun assault on civilian targets. Enforcement would eliminate a source of the passports and other official documents that terrorists have been using to cross national boundaries. Enforcement also would dry up at least some of the sources of the arms, explosives and ammunition used by such terrorist groups as the Irish Republican Army, the Red Brigade, the Shining Path and the Basque Fatherland and Freedom organization.

The law and its vigorous enforcement would strike a major blow against terrorism. But it would be nonsense to claim that all terrorism would be eliminated. In places like Lebanon—where government has been replaced by anarchy—murder, kidnapping and torture have become commonplace methods of gaining tribal advantage or preventing tribal annihilation. This pattern and the variations that can be seen in such locations as Africa, Afghanistan and Central America are unlikely to be changed in the near future. Even with the law and the enforcement agency in place, ethnic, religious and ideological struggles will continue to erupt from time to time and, in all likelihood, civilians will be subjected to unspeakable cruelties. Despite the law and the agency, freelance terrorist groups well may continue to operate in many parts of the world, taking advantage of the mobility of our age by moving from country to country and attempting to demonstrate their power and importance by striking at targets of opportunity— innocent and vulnerable civilians.

But while the number of terrorist attacks will not drop to zero, the frequency will be greatly reduced and so will the pervasive fear that is one of the most devastating results of terrorism. Thanks to the anti-terrorist law, business people no longer will be afraid to travel to London, Athens or Ankara. Tourists will able to visit the world's great cities without having to worry about a grenade attack on a sightseeing bus or an outdoor cafe. These very important, tangible benefits could help attract broad public support for the utilization of CDP in the development of an enforceable anti-terrorism law. Unlike the vast, institutionalized nuclear threat that tends to be forgotten or ignored, the threat of terrorism is specific and comprehensible. A practical plan for the reduction of this threat is bound to attract powerful public support and that support could result in a strong and persistent effort to apply the CDP concept to the terrorism problem.

Although CDP was developed to deal with the problem of weapons of mass destruction, it is entirely possible that its first application will be in the suppression of terrorism. If this happens, if the world's anti-terrorism effort makes use of CDP, I am convinced that it rapidly will demonstrate its effectiveness. It will give the world a nonviolent answer to a violent problem. It will set an example of how society can solve a problem involving violence of incomprehensible magnitude, the problem of weapons that have the power to reduce our planet to a cinder.

After living with these weapons for more than four decades, many people have desensitized themselves to the threat. They have closed their minds to the fact that our world is the target for 60,000 nuclear weapons and, if the nuclear warheads are launched, millions of us will be just 20 minutes away from incineration. Perhaps humanity is incapable of comprehending such a deadly possibility. Perhaps the key to building support for CDP is not more talk about the problem it will solve but about the world it can create. Perhaps we need to spend more time considering the other side of CDP—not what it can take away, but what it can give. And that's the terrain we will explore in the final section of this book.

# 14

# A Visit
# To The Age Of CDP

*When Kansas and Colorado have a quarrel over the water in the Arkansas River they don't call out the National Guard in each state and go to war over it. They bring suit in the Supreme Court of the United States and abide by the decision. There isn't a reason in the world why we cannot do that internationally.*

—Harry S. Truman

CDP will not create heaven on earth. It will not eradicate jealousy or hatred, poverty or disease. It will not guarantee social or economic justice and it will not eliminate all criminal behavior by individuals—or nations. But after CDP has been implemented, the world will be different and better. Trade will flourish. The exchange of information and ideas will expand. With the ratification of international law and the establishment of an appropriate monitoring and enforcement agency, the world will have in place a system that can banish the threat of armed aggression. Throughout history, whenever this threat has disappeared, economic benefits have followed. A prime example is the mutually beneficial relationship between the United States and Canada. Free of the fear of military action by its neighbor, each nation has reached across the border to establish useful ties without jeopardizing its national independence.

The same productive situation has been developing in Western Europe. With ancient threats vanishing into the past, nations that once were enemies have joined forces in the creation of a European Community that is destined to become a primary player in the world's economy. Europe's accomplishment, which was beyond imagination as recently as the end of World War II, will become a reality on a global scale with the implementation of CDP. Then people of all nations—developed or underdeveloped, socialist or capitalist—will reap abundant dividends and the most important of these will be psychological. For the first time in more than four decades, the shadow of the mushroom cloud will be lifted from human society and fear no longer will limit the potential of the human race. Citizens and their leaders will act with confidence in the certain knowledge that tomorrow will come and that tomorrow can be made better than today.

It is impossible to speculate about the magnitude of the psychological damage suffered during all the years of self-inflicted tension. There is no way to measure what happens in the mind when a feeling of helplessness forces it to deny the existence of a life-threatening condition. We do know what happens when individuals learn that they have a serious illness. The diagnosis itself inflicts a psychological wound. The victim is distracted by a new sense of vulnerability, a debilitating feeling of weakness. Even when there is no outward sign of illness, no physical disability, the challenge of routine responsibilities becomes increasingly burdensome. Sometimes it seems best to attempt to avoid the problem by denying that it exists. And this denial inevitably compounds the problem and makes recovery more difficult. If there is recovery, the experience has a profound effect on how a person views life. It becomes much more important. Each moment seems filled with meaning. Existence no longer is taken for granted. It is treasured and so are the families, friends and neighbors who share the precious gift of a life that has been restored.

---

**CDP insists that all nations have the right to defend themselves. CDP also maintains that all nations have the right to come to the assistance of nations that are the victims of aggression.**

---

When the world recovers from the madness of the present age, when enforceable international law has removed the malignancy of civilian-threatening weapons, we can expect a simliar kind of psychological renewal. Men and women will look to the future with optimism and they will be increasingly motivated to unlock the future's potential. With CDP, the cure for the sickness of aggressive international behavior will be lasting. Coming generations will have confidence that international security will be maintained despite inevitable changes in economic and political patterns. Unlike programs which call for total and universal disarmament, CDP insists that all nations have the right to defend themselves. CDP also maintains that all nations have the right to come to the assistance of nations that are the victims of aggression. This emphasis on defense gives permanence to international security under CDP, a permanence that cannot be achieved through total and universal disarmament.

Of course, it is attractive to speak of banishing all arms from the world. But in such a world, how will law-abiding nations protect themselves if an aggressor nation comes into existence? How will disarmament treaties ratified in the 1990s be enforced in the 2010s? How can the world's people be assured that aggressive behavior always will be repressed and that the perpetrators always will be punished? The CDP answer to these questions involves two interrelated levels of protection. On the first level is the defensive capability that nations will be able to maintain. The defense will be non-provocative. It will not threaten any neighbor. But it can be effective in protecting a nation and its people against a neighbor bent on territorial or ideological expansion. On the second level is the anti-weapons law and its enforcement mechanism. Because it is supported by a clear majority of nations and by nations clearly possessing overwhelming economic power, the law will be enforceable. The law will mobilize international authority and then utilize that authority to assist victims of aggression. Like a homeowner, a nation will be able both to defend itself and to count on help from the police.

Consider this possible chain of events. It is year number eight since the adoption of the anti-weapons law under the two-thirds ratification formula. Indeed, since the two-thirds majority was reached, other nations have joined in endorsing the law. The list of endorsers now includes all the members of the nuclear club, including those who managed to keep their membership secret. Also on the list are all the nations of Europe and North America and most of the nations of Africa, Asia and South America. After a somewhat uncertain beginning, the monitoring and enforcement agency is functioning smoothly. The debates over the definition of the weapons to be banned, the composition and voting procedures of the agency's legislative body and other issues, major and minor, have been resolved. Thanks to the efficiency of the agency's monitoring arm, skilled professionals are able to review and evaluate an unprecedented amount of thoroughly reliable information. The agency has its eyes and ears on the world, providing a continuing safeguard against potential lawbreakers.

The climate of international security is stimulating economic growth. Trade barriers erected during the years when fear ruled the earth are breaking down. For example, the United States is producing rare materials in the weightless environment provided by a Soviet space station. The Soviets, in turn, are buying U.S.-built supercomputers and hiring U.S. agricultural specialists to advise farm managers on how to increase productivity. The two superpowers—a term that largely has lost its meaning—also are competing for business in many parts of the world. But they are far from the only competitors. The nations of the European Community, Japan, Korea, Taiwan, China, Brazil, all the nations that have succeeded in making the transition into the age of industry and information, are vigorously striving to increase their share of the global market. The competition is creating jobs, expanding opportunities, fulfilling expectations.

Still, there continue to be areas of significant poverty, places where geographic and other conditions have conspired to block access to the earth's riches. In most of these places, international assistance projects are making some progress in improving standards of living. But the pace of improvement is slow, often because the crippling diseases of hatred and suspicion do not easily disappear from the minds of those who feel they have been oppressed or denied. It is in one of these places that the anti-weapons law undoubtedly will be tested for the first time. The population, goaded by feelings of resentment, the desire for ethnic or national recognition and the urge to quickly solve all problems, will turn to an unscrupulous leader. The leader then will set out to gain an advantage over the other nations of the region by obtaining or building an outlawed weapon, one that that can be used in a game of international blackmail, a weapon designed to frighten other nations into granting economic, territorial or other concessions.

The weapons construction project will not go undetected for long. The monitoring process has made great progress in the years that have passed since the 1988 signing of the INF Treaty. The agency's automated detection devices will gather information indicating that an illegal weapon very likely is being developed. The information will be carefully analyzed and then it will be presented to the agency's executive. Under the rules adopted by the agency, rules which are binding on all nations supporting the anti-weapons law, the executive will be required to verify the suspected violation through on-site inspection. A team of inspectors immediately will be dispatched to the nation that could be in violation. If they find no confirming evidence, the matter will be dropped. But if the site visit verifies the suspicions raised by the agency's surveillance devices, the executive will be required to develop specific recommendations and present them to the agency's legislative body. Of course, the suspect nation could refuse to admit the on-site inspection team. But under the law and its supporting rules, a refusal to admit representatives of the monitoring and enforcement agency is itself a violation that can lead to punitive action by the international community.

Let us suppose the evidence convinces the executive that an illegal weapons building project is underway. The executive most likely will recommend that an ultimatum be given to the offending nation: Immediately stop all illegal weapons work and destroy all of the materials and equipment used in this work. The executive also will recommend that all other nations be put on notice that they must stop trading with the offending nation. The agency's legislative body will be convened at once—possibly through an international, interactive television hookup—to review the evidence and act on the recommendations. In keeping with fundamental principles of justice, the suspect nation will be extended an opportunity to deny its guilt or explain its position. However, given the persuasive nature of the evidence, it is most unlikely that this opportunity will be used.

What is likely is an act of defiance by the suspect nation. Perhaps a neighboring state will be threatened with devastation unless it makes some major concession. Or perhaps an attack will be launched without any warning at all. Both possibilities are entirely in keeping with the irrational behavior of a leader who knowingly defies the international community by constructing a weapon the community has outlawed. If there is an act of defiance—a threat or an actual attack—it will not deter the agency from carrying out its mission. The agency's legislative body will vote overwhelmingly for the imposition of comprehensive sanctions against the offending nation. All trade will be banned and the offender will be isolated from the community of nations, separated from the economic mainstream, cut off from the goods and information that are essential to a modern society.

Of course, there may be a nation or two that will seek to undermine the sanctions by surreptitiously trading with the offender. Here again, the agency's surveillance capabilities will come into play. If illegal trade is detected, the executive soon will know about it and, with the approval of the legislative body, total sanctions will be extended to all who defy the anti-weapons law by trading with the offender. Nations identified as being involved in illegal trade will be able to appeal the sanctions to the agency's judicial arm. Such an appeal might be based on a claim of mistaken identify or the contention that the trade was conducted by private citizens without the knowledge or consent of governmental authorities. In any event, the international court will hear the arguments, study the evidence and hand down a decision. While the monitoring and enforcement agency is doing its work, CDP's other layer of protection will be playing its role. This layer involves the defensive capabilities that all nations will be able to maintain. The nation that has been victimized by its outlaw neighbor will be doing its best to defend itself with the troops, weapons, fortifications and strategies that are permissible under the law forbidding weapons of mass destruction.

The defensive efforts of the victimized nation well may prove to be inadequate. But immediate assistance certainly will be provided by nations that have accepted the rule of law. These nations also can be expected to quickly deliver humanitarian and other assistance if there is a nuclear strike or some other horrifying act of aggression. Such an act cannot be ruled out. But in the world of CDP, an attack will be countered by the swift and certain application of the power of withholding. There may be devastation of unimaginable scope. The world may suffer a serious wound. But as I have pointed out, the wound will not be fatal. The earth will survive and, in the end, the attacker will be forced to obey a law designed to preserve civilized society. An attack organized by a fanatic leader would be a tragedy of incalculable proportions. But it certainly would be a lesser tragedy than the one we face today.

Of course, the goal of CDP is to eliminate all acts of aggression, to protect all nations from violence. To achieve this goal, the monitoring and enforcement agency may strive to develop new defensive techniques involving technologies that presently are beyond the capabilities of the scientists and engineers. For example, the agency eventually may be able to equip itself with satellite-based lasers or other instruments that will be able to destroy weapons of mass destruction before they can be utilized. Such instruments would provide the ultimate defense against overt aggression by an outlaw nation. They would minimize the possibility of civilian slaughter, assure protection for even the weakest of countries and make the threat of total economic isolation even more effective.  With the deployment of these defensive instruments, the window of peril would be slammed shut.

The ability to destroy outlawed weapons from a distance may not be achieved for some time.  But when this ability becomes available, it must be subject to the same rules that will apply to the detection of weapons.  The agency created to monitor and enforce the international anti-weapons law always must be under, not above, the law.  As Dr. Burns H. Weston, professor of law at the University of Iowa, noted in an article in the Summer 1987 issue of Case Western University's *Journal of International Law*:

> If the history of the last 40-odd years has taught us anything, it is that there is little hope for genuine security, national or global, without a strengthening of the legal foundations for the non-military resolution of international disputes.  Even if other countries do not always follow suit, surely our country and our children's future will be better served if we try hard to build as peaceful and just a world society as we can—and while we still have a chance to do so.  To put it more bluntly, it is respect (or lack of respect) for international law that in the end will determine the fate of the Earth.  Rededication to the world rule of law and cooperation in this Age of Nuclear Anxiety is not a matter of choice.  It is, quite simply, a matter of survival.

With the implementation of CDP, the enactment of international law and the establishment of a monitoring and enforcement agency, there is hope not only of survival, but of a world in which humanity will climb to new heights of accomplishment.

# PREMISE V

## Reaping
## The Harvest Of Security
## Achieved Through
## Enforceable International
## Law

Support for international law would bring to a nation all the economic, technical, social and other benefits of world trade. Violators of international law and their accomplices would forfeit these benefits and be subjected to complete economic isolation and worldwide public scorn.

# 15

# Winning U.S. Hearts
# And Minds

*An army of principles will penetrate where an army of
soldiers cannot.*

—Thomas Paine

In many ways, Premise V is a kind of summing up—a thumb-
nail description of the rewards and punishments built into CDP.
The description is direct and factual. Nations supporting the anti-
weapons law stand to gain benefits of enormous significance; violators face severe penalties. It is a concept that can change the
world, one that deserves the support of men and women of goodwill in every nation. Yet, hope for the realization of the CDP
concept rests most heavily on a single segment of the world's
population—thinking, caring U.S. citizens and like-minded citizens
of other nations.

Now I am not making any claim that Americans are blessed
with some extraordinary gift of intelligence or virtue. They have
not received any mystical anointing that specially qualifies them to
save the world from itself. In fact, they are not fundamentally
different from the people of any other nation. But there is one
thing that sets them apart: They control the destiny of a nation
that has enormous economic resources, a nation that must join
with its trading partners in providing the leadership that will bring
CDP to reality, establishing enforceable international law and
inaugurating a new age of lasting international security. If the
American people understand what CDP offers them and their

119

world, if they insist on positive action by their government, they can play an important role in creating a new and constructive era of international cooperation. They can join with the concerned citizens of other nations in bestowing on the world the gift that has eluded past generations—lasting security under the rule of law.

But building understanding among Americans will not be easy. CDP involves an unconventional approach to international affairs. It calls for a radical departure from current U.S. policy. It threatens the perceived interests of a number of groups within American society. If CDP is to be given the serious consideration it deserves, the concerns of all groups must be addressed. Americans in many walks of life must come to recognize the potential of CDP. Then, and only then, will U.S. leaders be willing to explore a new path to security—one that can make freedom from fear a reality.

It is impossible to overstate the importance of public opinion in the shaping of American policy. It surely and continually determines the course of action taken by those selected for governmental leadership. These public officials run serious risks if their actions are not in conformity with the thinking of a substantial number of their constituents. U.S. policy depends on the support of citizens acting as individuals or as members of interest groups. U.S. policy changes only in response to the concerns and hopes of those it is supposed to serve. In the international arena, public opinion will determine the shape and direction of the leadership the United States must help to exert. This leadership—absolutely essential for the implementation of CDP— will decide if the nations of the world are to enjoy the benefits outlined in Premise V. As British historian James Bryce put in his classic study *The American Commonwealth* (G.W. Putnam's Sons, New York, New York, 1959).

> Towering over Presidents and State Governors, over Congress and State Legislatures, over conventions and the vast machinery of party, public opinion stands out, in the United States, as the great source of Power, the master of servants who tremble before it.

On matters of foreign policy, public opinion often sends mixed signals to the leaders of the American nation. Everett C. Ladd, senior editor of *Public Opinion* magazine and executive director of the Roper Center for Public Opinion Research at the University of Connecticut, has listed some of the differences that

". . . exist side by side in a state of some tension."

> We want peace—and we want to see the spread of communism resisted. We want to see checks put on the arms race—and we want a strong defense, second to none. We want to sit down and talk with the leadership of the Soviet Union, and we always endorse the holding of summits—and we are deeply suspicious of the Soviet Union, doubting, for instance, that it will abide by its agreements.

In holding these diverse but not necessarily incompatible views, "Americans reflect," Dr. Ladd points out, "a complex, multi-sided set of foreign policy wishes and perspectives which in turn reflect the complexity of the world in which they live." In this complex world, the supremacy of public opinion is both a challenge and an opportunity for America. It is a challenge because the diversity of views necessitates the gradual evolution of policy changes. Abrupt alterations of course are beyond the power of U.S. leaders. On the other hand, the supremacy of public opinion constitutes an opportunity for America. In an opinion-dominated society there is room for new points of view on critical issues, new beliefs about the direction the world must take to assure its survival. In a society where public opinion is king, there is hope of reconciling differences, of creating a consciousness of the problems faced by humanity and a consensus on a solution to those problems.

---

**. . . by accepting the principle of peace through international law and non-violent enforcement of that law, the United States will be helping to launch an irresistible global movement.**

---

By moving in the direction of CDP, by accepting the principle of peace through international law and non-violent enforcement of that law, the United States will be helping to launch an irresistible global movement. It will be joining with other nations in taking the steps that will lead to the creation of an environment hostile to aggression, a climate that will foster the peaceful development of all lands, the security of all nations. Taking the lead will not be without its strains. Particularly affected will be those segments of the American economy that depend on defense production. As the building of weapons of mass destruction winds down, defense contractors and workers will have to adapt to changed conditions. They will have to adjust themselves to the demands and the competitive forces of a civilian-oriented economy. For some, this will not be easy. Years of reliance on the Pentagon procurement system are not likely to stimulate either innovation or an awareness of the requirements of market-place competition. If major contractors feel the squeeze, so will their sub-contractors. Dozens of relatively small builders of weapons components will be forced to switch to civilian production or go out of business. The impact will be especially evident in the 10 states—most of them in the Far West or South—in which nearly 60 percent of the defense budget is spent.

The dislocation also will be felt in academia. Colleges and universities are heavily involved in defense-related research and development. A reduction in weapons research will necessitate adaptation and perhaps shrinkage for at least some campus laboratories. It has been estimated that for every billion dollars cut from military spending, 35,000 jobs are lost. If this figure is accurate—or even close to accurate—the attempt to implement the Civilized Defense Plan most certainly will face powerful, well-organized and heavily financed opposition from what President Dwight D. Eisenhower described as "the military-industrial complex." The Pentagon and some of the other interests involved in the weapons production business will mount persuasive argu-ments against CDP. They will conjure up visions of a Soviet take-over of Western Europe and the Middle East. They will claim that the freedom of America is itself in jeopardy. They will denounce supporters of CDP by questioning both their patriotism and their intelligence.

The Pentagon also will claim that their deadend business makes a major positive contribution to America's economic well-being. They will argue that the production of weapons creates jobs and that the spin-off from weapons research contributes to the development of new or improved civilian products. These claims by the pro-weapons lobby have been skillfully developed and art-fully communicated since the end of World War II. They have been accepted by countless private citizens and endorsed by suc-cessive occupants of the White House as well as by the leadership of the congressional establishment. But despite popular accept-ance, the pro-weapons claims cannot stand careful analysis. Instead of strengthening the U.S. economy, weapons research and production has contributed to the weakening of some of the basic industries that long were the driving force behind U.S. prosperity. Manufacturers of steel, automobiles, machinery and electronic products have found it increasingly difficult to compete in the global marketplace. Newcomers to industrialization—Korea and Brazil, for example—have been chipping away at market segments once dominated by U.S. companies and, ironically, the United States also has been losing ground in the economic race to Germany and Japan, two nations thoroughly familiar with the consequences of military dominance of the economy.

In the United States, military considerations have created a *de facto* industrial policy that emphasizes high technology at the expense of basic industry. According to the Center for Economic Conversion, more than 70 percent of all federal spending for research and development has been for military purposes and, as a result, much of the nation's scientific and engineering talent has been involved in work for the aerospace, communications and related high tech industries. Meanwhile, basic industries have withered as leadership in production and product innovation has shifted from America to its overseas economic rivals.

Lloyd J. Dumas, professor of political economy and economics at the University of Texas at Dallas, points out that the military and military-related industries absorb huge amounts of productive resources. In his book, *The Overburdened Economy* (University of California Press, Berkeley and Los Angeles, California, 1986), Dr. Dumas notes that the United States and the Soviet Union ". . . have together accounted for half to two-thirds of worldwide military spending in recent years. Of course, it is not the number of dollars that is of real moment, but rather the enormous quantities of productive resources diverted from potentially contributive economic activity, which the dollars represent."

This diversion, Dumas makes clear, has had a serious negative impact on the economies of the superpowers, weakening their ability to compete in an increasingly competitive world economy. In the United States, it has been estimated that more than a third of all engineers, 25 percent of all physicists, 30 percent of all mathematicians and 20 percent of all machinists now work on military products. They do not, it should be added, devote much of their time to improving productivity. Production efficiency, a matter of vital importance in the commercial marketplace, is given minimal attention in the cost-plus environment of U.S. weapons design and manufacture.

To defend itself against criticism, the military and military-related industries sometimes make the claim that, over the years, military research has led to the development of some important civilian products. However, when you consider the scope of U.S. investment in military research, the number of civilian product breakthroughs is less than impressive. Indeed, it would be hard to think of a less efficient mechanism for utilizing research expenditures to create better products for civilian consumption. As weaponry grows increasingly sophisticated, the proportion of civilian breakthroughs to total expenditures is likely to decrease. At the same time, high tech industries will become more and more dependent on appropriations from the federal government. This dependency—perhaps addiction would be a better word— long has been recognized and exploited by members of Congress. They fight to get military research and production contracts for their districts. Congressmen realize that the loss of military business can mean increased unemployment and, quite probably, a hostile electorate.

Fortunately, dependency can be cured. In implementing the Civilized Defense Plan, research and production relating to weapons of mass destruction would be phased out gradually and, for the first three to five years, funds that had been earmarked for weapons of mass destruction would be used to assist industries converting from military to civilian production, a Marshall Plan approach for America itself. The objective would be twofold: to minimize economic disruption by making conversion as painless as possible and to prepare industries and workers for the role they must play in the enrichment of civilized society.

Think of the technological breakthroughs that could be achieved by redirecting the brainpower now devoted to the creation of frightful new weapons. These breakthroughs in such areas as robotics, transportation, generation and transmission of energy and genetic engineering would improve the quality of human life and, at the same time, create profitable new opportunities for business. Companies moving to take advantage of these opportunities by converting from military to civilian production could expect to prosper. Workers, managers and shareholders would benefit and, as reliance on Pentagon contracts diminishes, innovation and productivity once again would be hallmarks of American industry.

Conversion is a thoroughly achievable goal. In his book, *The Overburdened Economy*, Professor Dumas has provided what he calls model specifications for a National Economic Adjustment Act. The Act offers a detailed blueprint for a conversion process that could be achieved with minimal economic dislocation. It provides for job retraining, relocation assistance and other help for laid-off weapons workers. The financing of these benefits would come from a trust fund created with mandatory contributions from weapons contractors. To avoid the build-up of an expensive federal bureaucracy, the Act calls for the conversion planning to be done at the local level by a committee that would include representatives of management, labor and the community.

Alternative conversion plans already have been drafted for specific communities and industries and institutions such as the Center for Economic Conversion were established to offer practical, realistic and concrete advice on conversion issues. Thoughtful consideration of existing, carefully planned blueprints for conversion would relieve the fears of those who now are dependent on the development and production of weapons of aggression. It would underscore the fact that the Civilized Defense Plan is based on a common sense view of the situation now facing America and the world. In countering all the arguments of the weapons lobby, advocates of CDP will have common sense on their side. It is evident that the nuclear peril and the threat of chemical/biological warfare are too real and too sinister to permit the indefinite extension of the status quo. Fear of massive retaliation constitutes a frighteningly shaky foundation for the safety of the world, a fragile underpinning for the future of all of us. The conversion of the weapons industry to civilian production will seem destructive to those who are directly affected. But over the long term, even they will come to the realization that it is creative destruction that will give birth to new hope for humanity.

In his 1951 address to Congress, General Douglas A. MacArthur said, "I know war as few other men living know it, and nothing to me is more revolting. I have long advocated its complete abolition, as its very destructiveness on both friend and foe has rendered it useless as a method of settling international disputes."

In his book, *The Pathology of Power* (W. W. Norton & Co., New York, New York, 1987) Norman Cousins recalled a conversation he had with General MacArthur during the American occupation of Japan after World War II. The general said he felt the principal danger to the world's people stemmed from the failure to realize that military security, as the world had known it, was no longer possible and that world law had to be developed. Cousins notes that these thoughts were further developed in a speech by MacArthur to an American Legion group in 1955. The general described the horrors of modern war—"If you lose, you are annihilated. If you win, you stand only to lose." Then he went on to ask:

Must we fight again before we learn?  When will some great figure in power have sufficient imagination and moral courage to translate this universal wish (for peace)—which is rapidly becoming an universal necessity—into actuality?  We are in a new era.  The old methods and solutions no longer suffice . . . We must break out of the strait-jacket of the past.  There must always be one to lead, and we should be that one.  We should now proclaim our readiness to abolish war in concert with the great powers of the world.  The result would be magical.

MacArthur was not urging America to lay down its arms and put itself at the mercy of an aggressor.  He was not calling for peace at any price.  He was offering advice based on informed analysis and incomparable experience.  It is advice that echoes today in the thoughts and words of those who believe the Civilized Defense Plan offers new hope  for genuine peace.

Americans who support CDP believe that the United States— and every other nation under the sun—has the right to defend itself against aggression.  They believe in the maintenance of a strong defense.  They also believe in the necessity of international cooperation to eliminate once and for all the curse of weapons of mass destruction and the other immediate threats facing civilized society.  Advocates of CDP are not naive about the trustworthiness of the Soviet Union or other nations.  Promises can be broken.  Traditional territorial or doctrinal goals cannot easily be discarded.  A U.S. move toward implementation of the Civilized Defense Plan well may increase global suspicion.  For a period of time, CDP actually may seem to escalate international tensions.

In the face of this tension, the United States and the nations cooperating with it must be steadfast in their efforts.  They must continue to implement the Plan.  To back down, to return to the precarious security of the nuclear threat, would be a disaster beyond calculation.  It would condemn society to decades of peril. It would be a death warrant that might never be rescinded. Through resolute support of the Plan, the United States and its trading partners from around the world can shift the focus of international competition away from weapons of mass destruction and toward the weapons of trade.  The failed approach of armed aggression can be replaced by conformity with economic reality. Obsolete ambitions for political dominance on a regional or global scale can be assigned to the trashcan of history.  In their place will arise new concepts of competition through commerce, a new world of achievement in which every nation will be able to share.

# 16

# The Time Is Now

*There is a tide in the affairs of men,*
*Which, taken at the flood, leads on to fortune;*
*Omitted, all the voyage of their life*
*Is bound in shallows and in miseries.*
*On such a full sea are we now afloat,*
*And we must take the current when it serves,*
*Or lose our ventures.*

— William Shakespeare

When people look back from the vantage point of the future—
if, indeed, there is a future—they will see our age as a turning
point in human history. They will recognize that our time was
marked by the convergence of powerful forces, by changes destined to forever alter the relationships of nations to each other
and to our world. Historians will note that in the final decades of
the 20th Century, the obsolescence of military power became
increasingly apparent and so did the growing importance of
international commerce. Humanity had created a global economy
held together by a global network of communication and transportation. For the first time, the interdependence of all nations and
all people was not merely a pious belief but a demonstrable fact.

The historians of the future will emphasize the unprecedented
opportunity that change presented to the people of our age. They
will record that trade and technology thrust into our hands a key
that we could use to release humanity from the age-old prison of
fear, a key that could open the door that leads to a new and
higher level of human existence. If we seize the opportunity, we
will be acclaimed by future generations. They will think of our
time on earth as a golden age, a period marked by an extraordinary flowering of wisdom, compassion and courage, a moment

**127**

in which humanity dismantled the ancient machinery of aggression and focused the world's creative energy on the renewal of the earth and its people.

> **The historians of the future will emphasize the unprecedented opportunity that change presented to the people of our age.**

If we do not act, if we ignore these moments of opportunity, the generations of the future will view us with contempt. They will brand us failures. They will see us as men and women whose inaction condemned society to an eternity of hopelessness. Worse yet, if we do not act, there may not be any future generations of the human race. There may be no one on earth to hand down a verdict on our action or inaction. So why is there any doubt about what we will do? Why is there the slightest uncertainty about the direction we will take? With such priceless opportunities in our grasp and with the fate of the world in our hands, how could anyone question our resolve? Is it because the barriers to lasting international security seem insurmountable? Or because it often is easy for us to see obstacles and difficult for us to recognize the ways in which obstacles can be overcome? Or do we feel comfortable in accepting a status quo filled with incalculable danger for ourselves and our children?

Perhaps we have lived with violence so long that we are afraid to live in a non-violent world. Perhaps our constant exposure to violence has desensitized us to the compelling power of non-violence, to the constructive influences that are the foundation of civilized society. Perhaps we have been too busy with our own affairs to see what trade has accomplished or to understand how it shapes every aspect of our lives. Perhaps our tendency to glorify armed force has kept us from understanding that, more often than not, it has exerted a negative influence on the progress of civilization. Perhaps our problem is simply inertia. We expect someone else to take the initiative. We look to government to provide leadership. We convince ourselves that public officials have all the necessary information and expertise, that they will know when the time is right to move for the establishment of enforceable world law. We find it soothing to put out of our minds the harsh reality that a government's primary concern is the perpetuation of its own power. We choose not to trouble ourselves by recognizing that a government will take no steps to diminish its power—including the power to wage war—until it perceives that such steps will assure its own survival.

Perhaps the doubts about our willingness to act can be blamed on the selfishness that often seems an integral part of human nature. Selfishness prevents us from taking a long view of life. It blinds us to our responsibilities to others, including all those who have yet to breathe the sweet air of this precious planet. Selfishness keeps us from asking fundamental questions about our own existence. Why were we given the gift of life? Were we created to live out our days in isolation from our brothers and sisters? Or do we have an obligation to them? Have we been put on earth solely to satisfy our hunger for wealth, power or affection? Or to be contributing members of the human race? Do we have a responsibility only to ourselves? Or are we called on to pay the debt we owe to those who have gone before us, to make the world a better place for those who will follow?

We do have the power to lead the world into a tomorrow of unimaginable promise. We do possess the muscle to lift the world out of its rut and into a new path to international security. With all my heart, I believe that an individual or a group of individuals can achieve any reasonable goal. I believe we can make our dreams come true. But we must do more than dream. We must act. We must dedicate ourselves to the establishment of a world in which the rule of law is supreme. In such a world, every part of the global community will benefit. The United States, the Soviet Union and their allies will be able to lay down the armament burden and move forward with the peaceful conquest of more of humanity's unexplored frontiers. Developing nations will be free to concentrate on meeting the basic needs of their people, on providing food, shelter, education and health care through expanded economic opportunity. All nations will be able to take advantage of humanity's rapidly expanding base of scientific knowledge and technological achievement. This will be the world of CDP, a world that will leave behind the failed traditions of negotiations, treaties and geopolitical maneuvering, a world in which cynicism and despair will be replaced by idealism and hope.

Americans will not be strangers to such a world. They built a nation on a foundation of hope and for decades idealism was the controlling factor in U.S. relationships with friends and enemies. We were outspoken in our support of the beliefs that are at the core of the American experience—the rule of law, the peaceful settlement of disputes and the benefits of economic competition.

Now Americans must renew their commitment to these beliefs. We the people of the United States must stand up for what we know is right. Then we will be helping not just ourselves but all others. Even if there are disappointments and setbacks, we and our friends from around the world will be making progress toward a goal that must be reached.

What can we do? We can begin by bringing the CDP concept—the idea that economic power can establish true security through enforceable international law—to the attention of others. We can tell our co-workers and professional colleagues, our employees or employers, our clients and customers, friends and neighbors. We can tell our religious ministers and our local public officials. And we must make sure we communicate the CDP message to the members of our own families. Business people can play an especially important role. In today's economy, they have contacts across the nation and around the world. They can utilize these contacts to describe the CDP concept to an audience that should be receptive, an audience that should understand the realities of economic power and the interdependency of all nations and all people. The goal is to create a climate that is receptive to CDP, to gain the support of a growing segment of public opinion, one that crosses political and ideological boundaries. When this goal is reached, support for the CDP concept will rapidly gain momentum, capturing the attention of the policymakers in the White House and the Congress and moving America toward the position of cooperative international leadership that it has a responsibility to assume.

The task is great but far from impossible. As Willis W. Harman, president of the Institute for Noetic Sciences and a respected authority on social change, has noted, "Fundamental change in society has always come from vast numbers of people changing their minds just a little." By changing the minds of a few people with whom we have contact, we can create a ripple of change that will grow into a tidal wave. We can, quite literally, change the world. By striving to achieve positive change, we will be following one of America's oldest and strongest traditions. At the heart of this tradition is a spirit of optimism, a belief that citizens have the power to alter the course of history. This belief certainly was at work 200 years ago when 55 delegates met in Philadelphia to shore up the tottering federation of 13 former British colonies. The colonies were different in size, resources and outlook. The representatives of each colony were wary of any proposal that in any way threatened to lessen the right of a colony to go its own way.

Yet, out of the suspicion and jealousy of this Constitutional Convention came a document that is widely regarded as one of mankind's most remarkable political achievements. The Constitution did not eliminate the differences among the 13. It did create a framework of law in which the 13 could work together in the pursuit of important common goals. It was a framework necessitated by reality, by the realization that none of the 13 was strong enough by itself to independently assure its own survival. The parallel with the Civilized Defense Plan is clear. Like the U.S. Constitution, CDP is a response to a threat. In the case of CDP, the threat involves not the future of Georgia or Rhode Island or Delaware, but the survival of the human race. CDP, like the Constitution, accepts the differences that exist among political entities. CDP only asks a nation to surrender a fraction of its *perceived* sovereignty by supporting an enforceable international law banning weapons of mass destruction.

CDP does not seek to create a world government or to weaken the independence of any nation. On the contrary, its goal is to eliminate the weapons which are the greatest enemy of every nation's independence. The Civilized Defense Plan is not a utopian dream. It is a sound program that can liberate the world from nuclear and other civilian threatening weapons. It can protect against aggression, including the aggression of terrorism. It is a liberation movement and, like liberation movements of the past, it can change the world. Grassroots movements successfully abolished slavery and promoted such now fully acceptable causes as women's suffrage, workers' rights, religious tolerance and racial equality. The movements started with a tiny minority of citizens who endured pervasive public ridicule. But the minority persisted until it forged an effective consensus that forever transformed society.

In his long ago speech to Congress, President Roosevelt said the elimination of weapons of aggression was ". . . no vision of a distant millenium. It has a definite basis for a kind of world attainable in our time and generation." The goal, he said, is a "world order in which countries work together in a civilized society." You can be a partner in creating a civilized society. You can help free humanity from the fear of annihilation. By joining with all those who care about the world and its people, by demanding that international security be based on life protecting law instead of death dealing arms, you can bring about a new era of human achievement.

The time is right. The forces of change are at work on every continent. The opportunity for humanity to achieve a higher level of existence never has been greater. The dream of creating a world free of aggression never has been so close to realization. But unless you act, the dream will remain just that. Unless you speak out, the opportunity will be missed. Unless you invest your time, energy and, yes, prayer, this moment of hope will fade and civilized society once again will be condemned to an eternity of fear. By asking others to join you, by seeking to build grassroots support, by knocking on the doors of those who control our society's influential institutions—government, the mass media, education—you can assure that tomorrow will dawn for you and your children. You can achieve the most important objective to which any of us can aspire—the preservation of life on this wondrous planet that is our home.

> **The opportunity for humanity to achieve a higher level of existence never has been greater.**

# Afterword

Now that you have read *The Civilized Defense Plan*, you may be saying to yourself, "Nice idea, but too simplistic, too idealistic." If so, you are not alone.

All of us—and especially the most experienced and the most learned—are skeptical of concepts that challenge accepted practice and thought. This is normal. This is the way it has always been.

Ten years ago, I was one of the skeptics when it came to new plans for ending the arms race. This in spite of the fact that nuclear arms and our confrontation with the Soviet Union had been my greatest concern, next to my family and my business, for the previous 30 years.

Too simplistic? Is it possible for anything that can deliver the desired results to be too simple? I have never found it so, but I have found entanglement in compounded complexity to be an insurmountable obstacle. The few times in my life that I have achieved results beyond all expectations were when I was able to persuade the technicians and the other experts to forget the complexity of the problem and concentrate on concept.

With the concept clearly in mind and compounded complexity out of mind, I found things began to happen. With old molds of thought broken, I have seen new worlds of thought emerge. What once seemed impossible, through a different approach, was clearly possible.

Let me cite a simple example. In the early years of one of the businesses I established, we were working on a mechanism for automatically feeding poultry. I had stubbornly insisted that we

133

should make something different and better than anything on the market.

We were concentrating on using an auger for conveying feed. Though an auger did the job, it had a lot of disadvantages. It had to be spliced every 10 feet. It was difficult to make, handle and ship. One evening I said to a man named Eldon Hostetler, who had been working on this project, "We've got to solve this problem. We have to make it so that manufacturing, shipping and installation will be easier. Although I think it impossible, it would be ideal if we could make it in one piece." Then I walked away and promptly forgot what I had said.

But the next morning Eldon Hostetler came to me, excited as a little boy. He had a bucket full of feed. In it he had a tube containing what looked like a flattened wire spring. Turning the spring, he cranked the feed out of the bucket into another bucket. What he had done was take the central shaft out of the auger and put the strength of the auger in the flighting, the part that actually conveys the feed. Without the shaft, the auger could be made in one piece in any length desired. It could be coiled for easy handling and shipping. Because it was flexible, it could go around corners, something ordinary augers can't do. In use, one piece could do the work of the hundreds of pieces required by conventional augers and chain-conveying mechanisms. Today the centerless, flexible auger is the preferred way of conveying feed and other bulk materials in every developed nation and many developing nations. Thousands of miles of it have been made, probably enough to encircle the earth.

Now remember, the auger is not a new invention. You can find inscriptions in stone that picture ancient Egyptians elevating water out of the Nile into irrigation ditches by cranking an auger inside a tube. But why did it take 4,000 years for someone to realize that an auger could be made without a central shaft? Was it an accident? Did it happen because I said we had to find a better way? Or was it Eldon Hostetler's ability to think in simple terms, to strengthen what was essential and to eliminate what was unnecesary?

Undoing the Gordian knot of incomprehensible complexity, making form follow function, does appear simplistic to many. The act of simplification seems so obvious—after it is completed—that the normal comment is, "Why didn't somebody do this long ago?" But for 4,000 years nobody did, and I think I know why: Simplicity is not a simple matter.

As writers, artists, engineers and scientists know, reducing complexity to simplicity is the greatest of challenges. Complexity often is compounded by the accumulation of impulsive reactions and short-range surface thinking. In contrast, simplicity is a result of years of thoughtful incubation and cross-fertilization of ideas. History gives us countless examples of the power of the simplification process. But let us consider just one: Lincoln's Gettysburg Address. In its few words are compressed a sweeping vision of America's past and future, its problems and potential. It is a vision that reflects years of intensive study and profound thought refined in the crucible of fratricidal war. It is a vision that penetrates to the essence of what America is and what it must strive to be. I do not want to leave the impression that composing the Gettysburg Address or proposing a new approach to defense is identical to inventing a new kind of auger. But each of these developments reveals something about simplification. I chose to tell the story of the auger because of what it says about solving problems and developing concepts that work. It is analogous to the subject of this book in the sense that the solution to all problems is in the mind. This solution cannot be found until the problem is defined, a determination is made to solve it and a change of thinking is effected.

Changing the way we think is difficult. All of us are inclined to think of the present as a condition that will continue forever when in reality, and very fortunately, it never does. What we conceive of as problems are usually decisions waiting to be made, opportunities waiting to be discovered.

I have learned that I am much better at defining a problem and conceptualizing a solution if I don't get involved in technicalities. In the process of conceptualizing, I have found that it is a good idea to keep a respectable distance from people who are deeply involved with technicalties. Most of them are quick to tell you that what you would like to do is impossible. When working with many specialists, I think it's a good rule not to ask them what you can do, but to tell them what you want to do and challenge them to find a way to do it.

If this policy generally produces desired results, and I have found that it does, why not employ it in the foreign policy arena? Why not tell international experts and our government leaders what kind of a defense we want and challenge them to deliver it?

I'm convinced that those in control of our government, especially in the State Department and the Pentagon, are very much like other experts. They know more about what cannot be done than about what can be done through a different, perhaps unconventional approach. I am also convinced that until the American people speak with authority saying, "We've got to solve this problem. We must find a better way," we will not end the armaments impasse.

But what about the criticism that the Civilized Defense Plan is too idealistic? Is this a thoughtful criticism? An alibi for lazy thinking? Or an excuse for not thinking? My guess is that it's one of the latter two. For a number of reasons, people think only superficially or not at all about the international weapons problem.

Many feel there is nothing they can do about the problem so they shut it out of their minds. Others are afraid to think. The awesomeness of nuclear and other weapons of mass destruction literally paralizes the thinking of these people. Many others think of the problem as belonging to someone else and they feel no obligation to help solve it.

This kind of thinking, or refusal to think, has contributed to our present peril. It has built a roadblock that bars us from a more secure future.

We as individuals need to demolish the self-imposed barrier and put our minds and energies to work. We need to assume our clear responsiblity for directing our country. We need to say to ourselves, "This is my nation. This is my problem. This is my life that is being threatened. I must do something now to find a way to remove this threat to myself, my children, my hope of posterity."

As long as we refuse to think in these terms, as long as we shirk our duty to ourselves, our country and the world, the problem of earth-threatening armaments will remain.

When the concept of the Civilized Defense Plan burst into my consciousness in September, 1979, I thought of it as an answer to a subconscious wish which had been building over many years. I did not immediately understand it. As some critics of the Civilized Defense Plan say, "It was too simplistic, too much of a statement of the obvious to be taken seriously." But I soon found out that, like a new-born infant, it clamored for my attention and had a will of its own.

Experience made me realize that giving birth to or recognizing a concept was really no accomplishment. It was an involuntary response to accumulated thinking and it required no effort. Developing the concept into an instrument with the ability to achieve the desired outcome is quite another thing. It requires voluntary effort in enormous quantities and input from many sources.

The wisdom of people in academia, government, business, the sciences, the military, law and international relations, peace movements, religion and many other areas contributed to making the Civilized Defense Plan what it is. It was not until I received the counsel, encouragement and support of some very learned successful people that I fully comprehended the Plan's potential impact.

With others, as with myself, I've noticed that full comprehension of the Civilized Defense Plan takes time and thought. Like $E = mc^2$, CDP is not as simple as it looks.

Too idealistic? Only if we think the promise of the Fourth Freedom is more idealistic than the promise of freedom given to us by our forefathers.

They established the United States of America on the concept of a government of the people, by the people and for the people. It is a concept that is gaining ground in many nations, one we must revitalize here at home in the cause of international security.

The generation preceding us assumed the responsibility of defending us from fascism. Our generation has inherited the responsibility of eliminating weapons of mass destruction. We must rise to our defense, as did our fathers, to insure that generations yet to come will live free of the fear of annihilation, that they will enjoy the new age of peace that we can create.

Our position in history is unique. We hold in our hands the power to destroy all that we have inherited from God and our ancestors, to erase all hope of posterity. We also hold the key to a door that leads to a higher level of human existence, to the fulfillment of economic, aesthetic and spiritual aspirations, to an epoch of achievement that is beyond our imagination. We have reached a moment in time when our destiny depends on our will— or the lack of it.

We cannot ignore our own responsiblity. If the leaders of the United States will not lead, we the people—you and I—must fill the leadership vacuum. We must march forward toward our vision of the future and, when we do, our leaders will follow.

By daring to be pioneers in our own age, we can bring the Fourth Freedom to reality. We can create a world free of the fear of destruction, ready for the endless opportunities that wait beyond the threshold of tomorrow.

We have the power to give to our children a priceless gift, a gift worth more than all our money and all our possessions—the gift of life.

# About
# The Fourth Freedom Forum

The Fourth Freedom Forum was founded by Howard S. Brembeck in 1982 to help lead America and the world toward the goal of the Fourth Freedom—Freedom from Fear. Through an active outreach program, the Forum stimulates informed public discussion of international seciruty issues, giving special emphasis to the utilization of economic power to eliminate aggression from the face of the earth.

While the Forum believes economic power must replace military power in the establishment and enforcement of international law, it also strives to exchange ideas with interest groups reflecting a variety of experiences and beliefs. These include business, the professions, educators, the communications industry, service clubs, and mainstream organizations. The objective of this dialogue is the development of broad international public support for security through economic power, a concept based on the realities of today's interconnected, interdependent world.

The Forum is constantly striving to develop innovative tools to educate and inform. These include an award-winning 30-minute video that offers a thought-provoking overview of the Civilized Defense Plan. The Forum also has a speakers' bureau and its newsletter, INFORUM, is distributed nationally and internationally.

The Forum is a nonprofit, nonpartisan and nonsectarian organization directed by a board that includes representatives of business, education, religion, the professions and other interests. Providing additional insights and advice is a group of advisers that includes distinguished specialists on international security issues. Membership in the Forum is open to all who desire to make the vision of the Fourth Freedom a reality for the people of all nations.

For additional information, contact:

The Fourth Freedom Forum
803 North Main Street
Goshen, Indiana 46526
1-800-233-6786
Fax 1-219-534-4937